WETLANDS
*Environmental Gradients,
Boundaries, and Buffers*

WETLANDS
Environmental Gradients, Boundaries, and Buffers

Edited by
George Mulamoottil
Barry G. Warner
Edward A. McBean

CRC Press
Taylor & Francis Group
Boca Raton   London   New York

CRC Press is an imprint of the
Taylor & Francis Group, an informa business

CRC Press
Taylor & Francis Group
6000 Broken Sound Parkway NW, Suite 300
Boca Raton, FL 33487-2742

© 1996 by Taylor & Francis Group, LLC
CRC Press is an imprint of Taylor & Francis Group, an Informa business

First issued in paperback 2019

No claim to original U.S. Government works

ISBN 13: 978-0-367-44861-5 (pbk)
ISBN 13: 978-1-56670-147-1 (hbk)

Visit the Taylor & Francis Web site at
http://www.taylorandfrancis.com

and the CRC Press Web site at
http://www.crcpress.com

Library of Congress Cataloging-in-Publication Data

Wetlands : environmental gradients, boundaries, and buffers :
    proceedings of an international symposium, April 22-23, 1994 /
    sponsored by the Wetlands Research Centre, University of Waterloo,
    Ontario, Canada ; edited by George Mulamoottil, Barry G. Warner,
    Edward A. McBean.
        p.   cm.
    Includes bibliographical references and index.
    ISBN 1-56670-147-3 (alk paper)
        1. Wetlands—Congresses. I. Mulamoottil, George. II. Warner,
Barry G. III. McBean, Edward A. IV. University of Waterloo.
Wetlands Research Centre..
    GB621.W47   1996
    551.4'57—dc20
                                                                95-46404
                                                                CIP

Library of Congress Card Number 95-46404

PREFACE

An understanding of environmental gradients (physical, chemical, hydrological, and biological) is a prerequisite to the delineation of wetland boundaries. Only when boundaries are determined can an appropriate buffer be established. With this in mind, this symposium set out two objectives: (a) to focus on current research on environmental gradients, boundaries, and buffers, and (b) to provide a forum for interested researchers to generate recommendations on wetland priority research needs related to the theme of the symposium.

There were 65 participants who attended the symposium over the two days. The importance of the symposium to wetland science, public policy, and practice attracted participants from all levels of government, nongovernmental agencies, universities, the consulting industry, and private groups. The papers included in this volume were presented during the symposium and represent a wide range of viewpoints held by academicians, environmentalists, policy makers, consultants, planners, engineers, hydrologists, biologists, geochemists, ecologists, and conservationists, all having a common interest in the understanding of wetland boundaries, buffers, and environmental gradients. A final section provides a summary of the main issues of concern raised and a number of recommendations arising from a panel discussion.

D. Bonner of the Environmental Studies Cartographic Centre and S. Chisholm and S. Marsters of the Wetlands Research Centre assisted with the preparation and organization of the symposium. S. Smith assisted with the manuscript preparation. The Wetlands Research Centre and the Faculties of Engineering, Environmental Studies, and Science at the University of Waterloo provided logistical and financial support. We thank all symposium participants, session chairs, and authors of papers for their contributions. Finally, we acknowledge Lewis Publishers and Ms. Andrea Demby for their patience and assistance in publishing the book.

George Mulamoottil
Barry G. Warner
Edward A. McBean

THE EDITORS

George Mulamoottil, Ph.D., is a Professor in the School of Urban and Regional Planning and the Department of Geography at the University of Waterloo. He is currently Associate Dean of Graduate Studies and Research in the Faculty of Environmental Studies. Dr. Mulamoottil's research interests have focused on the planning and management of water resources. His recent contributions are in the area of wetlands and stormwater management. Dr. Mulamoottil has worked as a civil servant and has been in the environmental consulting practice in Ontario for the last 25 years.

Barry G. Warner, Ph.D., is Associate Professor of Geography with cross-appointments to the Departments of Biology and Earth Sciences at the University of Waterloo. He has been a Post-doctoral Research Fellow at the University of Helsinki in Finland and a Research Fellow of the Swiss National Science Foundation at the University of Neuchatel in Switzerland. He is currently the Chair of the Canadian National Wetlands Group and a Council Member of the Canadian Society for Peat and Peatlands. His research focuses on wetland ecosystem dynamics and the causes and consequences of change with particular attention to peatlands. He has studied peatlands in Canada, the United States, Scandinavia, Europe, and Russia.

Edward A. McBean, Ph.D., is an Associate of the firm Conestoga-Rovers and Associates Ltd., an international environmental engineering company with headquarters in Waterloo, Ontario. He has held academic appointments at the University of Waterloo, University of California, and Cornell University. He has written one book, edited four books, and is the author of 140 articles in refereed technical engineering journals. Dr. McBean is a member of the Wetlands Research Centre at the University of Waterloo and has been involved in assessing the impacts of urbanization on surrounding land uses as it relates to water quantity and quality.

CONTRIBUTORS

C. W. P. M. Blom
Department of Ecology
University of Nijmegen
Toernooiveld
6525ED Nijmegen, The Netherlands

P. S. Botts
Division of Science
Pennsylvania State University – Erie
Erie, Pennsylvania 16563

G. Bowen
Ontario Ministry of the Environment
 and Energy
135 St. Clair Ave. W.
Toronto, Ontario, Canada M4V 1P5

Y. Z. Cao
Environment Canada
Canadian Centre for Inland Waters
New Technology Branch
867 Lakeshore Road
P.O. Box 5050
Burlington, Ontario, Canada L7R 4A6

V. Carter
U. S. Geological Survey
430 National Center
12201 Sunrise Valley Drive
Reston, Virginia, U.S.A. 22092

J. E. Cox
School of Urban and Regional
 Planning
University of Waterloo
Waterloo, Ontario, Canada N2L 3G1

R. Donn
Division of Science
Pennsylvania State University – Erie
Erie, Pennsylvania 16563

J. Z. Fraser
Ontario Ministry of Natural Resources
Southern Regional Office
50 Bloomington Road
Aurora, Ontario, Canada L4G 3G8

P. J. Harpley
South Lake Simcoe Naturalists
R.R. #1
Pefferlaw, Ontario, Canada L0E 1N0

A. B. F. Hinton
Ontario Ministry of Natural Resources
Grenville West/Dundas Area
Kemptville District
P.O. Box 605, Oxford Street
Brockville, Ontario, Canada

D. B. Hodgson
Ecologistics Ltd.
490 Dutton Dr., Suite A1
Waterloo, Ontario, Canada N2L 6H7

M. M. Holland
Department of Biology
The University of Mississippi
University, Mississippi
U.S.A. 38677

E. A. McBean
Conestoga-Rovers and Associates Ltd.
651 Colby Drive
Waterloo, Ontario, Canada N2V 1C2

K. J. McKague
Ecologistics Ltd.
490 Dutton Drive, Suite A1
Waterloo, Ontario, Canada N2L 6H7

J. M. McMullen
Terrestrial Environmental
  Specialists, Inc.
23 County Route 6, Suite A
Phoenix, New York, U.S.A. 13135

P. A. Meacham
Terrestrial Environmental
  Specialists, Inc.
23 County Route 6, Suite A
Phoenix, New York, U.S.A. 13135

R. J. Milne
Department of Geography
University of Guelph
Guelph, Ontario, Canada N1G 2W1

G. Mulamoottil
School of Urban and Regional
  Planning, Department of Geography
  and Wetlands Research Centre
University of Waterloo
Waterloo, Ontario, Canada N2L 3G1

A. J. Norman
Ontario Ministry of Natural Resources
Southern Region
50 Bloomington Road West
Aurora, Ontario, Canada L4G 3G8

Z. Novak
135 St. Clair Avenue W.
Ontario Ministry of the Environment
  and Energy
Toronto, Ontario, Canada M4V 1P5

W. G. Pearsell
BOVAR Environmental Waste
  Management Services
17320-106A Avenue
Edmonton, Alberta, Canada T5S 1E6

J. D. Phillips
Department of Geography
East Carolina University
Greenville, North Carolina, U.S.A.
  27858

K. Richardson
Ontario Ministry of Natural Resources
Niagara North Area
Cambridge District
P.O. Box 1070
Fonthill, Ontario, Canada L0S 1E0

D. Routly
Ontario Ministry of Natural Resources
Brant/Wentworth Area
Cambridge District
P.O. Box 2186, Beaverdale Road
Cambridge, Ontario, Canada

D. E. Stephenson
Biology Department
Ecologistics Ltd.
490 Dutton Drive, Suite A1
Waterloo, Ontario, Canada N2L 6H7

R. W. Tiner
U.S. Fish and Wildlife Service
300 Westgate Center Drive
Hadley, Massachusetts, U.S.A. 01035

J. Uusi-Kämppä
Institute of Soils and Environment
Agricultural Research Centre of Finland
FIN-31600 Jokioinen, Finland

H. M. van de Steeg
Department of Ecology
Toernooiveld University of Nijmegen
6525 ED Nijmegen, The Netherlands

L. A. C. J. Voesenek
Department of Ecology
Toernooiveld University of Nijmegen
6525ED Nijmegen, The Netherlands

B. G. Warner
Department of Geography and
  Wetlands Research Centre
University of Waterloo
Waterloo, Ontario, Canada N2L 3G1

T. Yläranta
Institute of Soils and Environment
Agricultural Research Centre of Finland
FIN-31600 Jokioinen, Finland

CONTENTS

1 INTRODUCTION

G. Mulamoottil, B. G. Warner, and E. A. McBean

Wetlands are generally thought to be among the most fertile and productive ecosystems of the world. They provide a variety of ecological functions to the landscape. In recent years there has been considerable research activity to generate more scientific documentation on the ecological functions. Much of this activity has resulted from the need to protect wetlands and their functions which is of serious concern in most industrialized countries and, in recent years, in developing countries. The reasons for the protection and conservation of wetlands have been summarized by Greeson et al., (1979) and Williams (1990). Suffice it to say that "civilization began around wetlands; today's civilization has every reason to leave them wet and wild" (Maltby, 1986).

The values attributed to wetlands vary from one human user to another. Wetlands can be drained and used for agriculture or developed into a subdivision for housing or other land uses. They have been used as receiving areas for waste discharges by some, while some naturalists consider wetlands as unique ecosystems which should be protected. According to Richardson, (1985) the multitude of values ascribed to wetlands dictate that managers must question how the specific objectives of, for example, agriculture, energy production, fishing, hunting and recreation, forestry, and scientific study can all be satisfied within the context of regional ecosystem integrity.

The importance and significance of the functions and values of wetlands within the general scheme of landscape ecology have been receiving increasing attention during the last two decades. The early focus of the research was to establish the extent of the disappearance of wetlands by human use. For example, the wetlands in the Great Lakes region in central North America is estimated to be 67% of the original wetlands that existed prior to the time of settlement (Podniesinski et al., 1995). In Canada, during the last 200 years approximately 14% of the wetlands have been changed to other land uses (Rubec, 1994). In comparison to the wetland loss across Canada, the original wetland area in Ontario has been reduced by 68% south of the Precambrian Shield. Further, the wetland loss has been most severe in southwestern Ontario, where over 90% of the original wetland has been changed to other land uses (Snell, 1987).

The decline in coastal marshes on the Great Lakes has been of particular concern (Lemay and Mulamoottil, 1984; Krieger *et al.*, 1992). In one case study, Lemay and Mulamoottil (1984) estimated that urban and urban-oriented uses were responsible for a 50% decrease in marsh area on the Toronto waterfront. With the ongoing loss of wetlands, it is important to realize that there is a loss of wetland functions. Several animal extinctions have occurred as a result of reduced wetland habitat on the waterfront as determined largely from bird and wildlife census data (Lemay and Mulamoottil, 1984).

The Ramsar Convention of 1971 on wetlands was the first international convention for the protection and conservation of a specific ecosystem or habitat type. Agreement was reached among the different countries to facilitate the conservation of biological diversity. The convention espoused conservation of wetlands of international significance through the principles of wise use. More recently, the Convention on Biological Diversity of 1993 has provided further opportunities for reinforcing the activities of the Ramsar Convention (*Ramsar Newsletter*, 1994).

A number of wetland features such as the plant and animal communities, size, connection to other wetlands or water bodies, and the amplitude of water level fluctuations are, to varying degrees, essential for the continuing viability of wetlands. However, the most important determinant for the establishment and maintenance of specific types of wetland and wetland processes is hydrology. Indeed, it is the main driving force. Thus, for example, alterations to the hydrological regime as a result of urbanization can have significant physical, chemical, and biological effects on the wetland environment and its ecological processes. Even small changes in surface and ground water hydrology may result in significant changes to the wetland. These changes in turn may have cumulative effects downstream. However, prediction of the changes of wetland as a response to changes in hydrological conditions has been difficult.

Of concern, then, is to improve our understanding of the viability of wetlands to changing environmental conditions. Any wetland is in a continuous state of evolution and flux and should, with time, change into a terrestrial ecosystem. Understanding the propensity for natural ecological change is further compounded by humans, who have changed and continue to change land uses in and around wetlands. Urbanization is probably the most dramatic example of such a change to the natural landscape. Although significant efforts have been directed towards the creation of wetlands in urban areas, it should be noted that many wetlands are not truly viable in an urban environment unless followed by intense management.

More and more suburban land use planning is integrating wetlands and other natural areas as important features of the urban landscape, especially for recreational activities. In certain instances the urban wetlands are used to assist with the control of stormwater runoff prior to discharge into receiving water bodies. It must be borne in mind that suburban landscapes often radically reshape the natural lay of the land and thereby alter the infiltration and runoff characteristics. With the changes in the quantities of surface runoff and infiltration to ground water, there can be changes to the nutrient fluxes into the wetlands. In addition to direct physical and hydrological alterations that result from the construction phase, there are often other physical, chemical, and biological effects that can impact wetlands.

On the positive side, wetlands function as temporary traps of contaminants arising from urban runoff. For example, the impact of highway runoff on water quality often results in changes in turbidity, pH, conductivity, and temperature of water and by adding pollutants such as heavy metals, grease, and oil, contribute to the pollution of the receiving waters. Contaminants such as heavy metals and nutrients, e.g., phosphorus, may be temporarily retained in wetland vegetation or sediments. The vegetation and sediment can be removed, but these activities result in further impacts on the physical and biochemical regime of the wetland. The attenuation capability of wetlands, by removing various chemicals arising from urbanization, explains, in part, their widespread use to protect receiving water bodies. It is important to identify the type of wetland ecotones that are most important for intercepting nutrients and sediments. The degree to which wetlands behave as nutrient and contaminant traps varies with the season. However, to what degree are these attenuative capabilities predictable and to what extent will the viability of a wetland remain in its natural state?

As part of the commitment to protect significant wetlands in Ontario, the provincial government has mandated that an environmental impact assessment of all development activities be completed to assure that the functions and values of wetlands are not reduced. The construction of transportation facilities and other utilities in and around wetlands also requires such an assessment. Although simple in concept, the carrying out of the assessment is a major challenge. For example, the prediction of impacts of bridges or earthfills on the ecology of a wetland is far from an exact science. The placing of fills on wetlands can have physical, chemical, and biological effects on the ecological processes of a wetland. The contaminants from highway runoff entering wetlands may undergo bioaccumulation and even biomagnification, depending on the food chain phenomena. Clearly, detailed predictions in response to changes in these types of inputs are nontrivial.

There are considerable development activities in and around wetlands. In rural situations, the development pressures are related to the intensity of agricultural operations. The demands in the urban context are from the housing, industrial, commercial, and infrastructure sectors. Regulatory jurisdictions and planning agencies are attempting to regulate land use in wetlands and also on lands adjacent to wetlands. However, there is considerable uncertainty, which makes predictions of the impacts of development on wetlands rather nebulous. For example, it is difficult to forecast the impact of a particular land use on the functions of a wetland. Part of the difficulty is a general lack of understanding of wetland science and information gaps in wetland ecosystem dynamics. The practitioners produce environmental planning reports on developments based on the best available literature in wetland science. In fact, most such reports go beyond what can reasonably be said considering the little experience gained to date from wetland-related developments. Here is a situation where the needs of professional practice are ahead of the fundamental scientific knowledge base.

The development decisions related to wetlands are further influenced by policies on wetlands and other regulations. A recent study by Stadel *et al.*, (1995) has demonstrated that there are uncertainties and inconsistencies in the application of policies. An interesting aspect that was pointed out by the authors was the number of inconsistencies in the quality and quantity of information supplied by the witnesses at the Ontario Municipal Board hearings. They attribute this to the lack of scientific knowledge on the ecological functions and processes of wetlands. A significant point to note is the development of public policies on wetlands planning and management without sound scientific knowledge.

The protection of wetlands from land use impacts requires the establishment of buffers. However, the efficacy of the buffers and their long-term viability are only beginning to be investigated. In such a situation, it is difficult for regulatory agencies to accept or reject development proposals involving the use of buffers with different characteristics. Further, it is interesting to note that the width of buffers is stipulated from the boundary of a wetland. With the current problems associated with the boundary delineation issue, it is difficult to agree on a generally acceptable boundary. Such a predicament makes development decisions involving wetlands a vexing issue. The need for this symposium was conceived in the backdrop of such concerns on the protection of wetlands and with an interest to explore the ability of wetlands to function as attenuative features.

The questions and concerns far exceed the range of issues that can be addressed in a symposium. Nevertheless, a number of papers that highlight many interesting issues in relation to gradients, boundaries, and buffers are contained in this volume. Some of the papers are prescriptive, and others are summaries of recent research findings related to specific examples. In dealing with these issues, the papers have been divided into three theme areas, namely: environmental gradients, boundaries, and buffers.

ENVIRONMENTAL GRADIENTS

Issues of environmental gradients relate to the rates of environmental changes with distance or proximity. For example, Phillips in Chapter 14 points out how stream power is reduced by infiltration rates and gentler slope gradients. This has implications in the ability of buffers to remove or treat specific pollutants.

Alternatively, Blom *et al.*, in Chapter 7 describe a study of adaptive mechanisms of wetland plant species occurring along floodplain gradients. These authors examine the diversity of plant communities in relation to the different conditions caused by transient flooding of river floodplains and adaptive responses during periods of submergence. They utilize a "one-species-one-habitat" approach in which various species growing under the same conditions in the field are compared. Life-cycle strategies that have developed in plants to survive the stress conditions during periods of inundation are often dependent on a rapid transition from an active life to a more passive "wait and see" behavior. A part of the strategy can involve the ability to flower and to produce seed immediately after conditions improve; however, the ability to predict when these conditions will exist is difficult.

A quantitative modeling approach developed by McBean *et al.*, in Chapter 5 indicates that attempts to provide pre-urbanization infiltration levels in postdevelopment urbanization scenarios are going to be extremely difficult. There will be significant costs in dollars and space if mitigation of the detrimental changes to the hydrologic balances are planned. It is especially difficult to maintain the gradients for ground water recharge toward wetlands if the historical ratios of surface and ground water gradient inputs to the wetlands are to be maintained. Their findings indicate that measures to enhance infiltration are going to be necessary, if there is an interest to sustain the predevelopment conditions. To achieve environmental sustainability, mitigation measures in the form of small-scale infiltration facilities should be spread throughout the contributing drainage areas.

Chapter 4, by Warner, is a reminder that wetlands, and peatlands specifically, are three-dimensional landforms. Gradients not only occur on the land surface across wetlands but also exist vertically through them. The vertical gradient is essentially a "time gradient" which can change physically, chemically, hydrologically, and biologically with age. By relating these features of the vertical gradient with time, it is possible to determine the condition of the peatland prior to European settlement and the rates and the direction of change as a consequence of surrounding land use activities. This may be an important approach for monitoring change and predicting the future condition of wetlands under specific management strategies. In densely settled regions such as southern Ontario, the peatland condition of 10, 50, or 100 years ago may already represent a human-altered ecosystem and so management may have to use the peatland condition at the time before European settlement as the datum for gauging the peatland condition today.

BOUNDARIES

The delineation of wetland boundaries is extremely difficult. The statement by Canny (1981) "do not stand on your dignity about the real existence of any boundary; it is in your mind", captures the uncertainties associated with the determination of boundaries. In the U.S. the delineation is carried out by considering hydrology, soils, and vegetation. Some authors utilize the presence of surface water and interstitial soil water within the major portion of the plant root zone during the growing season as an indicator of wetland hydrology, but this approach is not universal. In Chapter 2 Carter describes the criteria currently in use, the problems associated with the procedures, and the merits of a classification system for identification of wetlands.

Tiner, in Chapter 8, discusses alternative methods for identifying and delineating wetlands, with definitions of boundaries encompassing a wide array of "wetlands" including marshes, bogs, swamps, fens, pocosins, and wet meadows.

As he points out, most wetland definitions emphasize the presence and predominance of plants that grow in water or in periodically flooded or saturated soils.

Alternatively, in Chapter 11 Harpley and Milne argue that wetland delineation should consider the faunal component along with the botanical, hydrological, and geomorphological criteria. This broader view of wetland boundaries arises because animals may rely on a wetland for only a portion of their lives. Avian populations associated with wetlands and the mapping of representative wetland species should examine habitats and wildlife disturbance zones. These authors emphasize that attention needs to be paid to the avifauna in the delineation of wetland boundaries.

Pearsell and Mulamoottil utilize hydrologic processes in Chapter 9 to define a wetland boundary. The use of functional boundaries in land use planning is emphasized in their chapter. The delineation of boundaries for different wetland functions is essential in order to incorporate wetlands into new neighborhoods.

Holland describes in Chapter 3 how the boundaries between wetlands and other components of the landscape are important transition zones between upland and aquatic ecosystems. The chapter reviews current thinking on critical research needs and the need for environmental monitoring.

The concerns and issues in delineating wetland boundaries in specific field applications are described in two interesting ways. In Chapter 6 Stephenson and Hodgson describe the impact of shallow subsurface water flows in the root zone and the nature of land use changes. They describe the utilization of time domain reflectometry and soil tensiometers to estimate volumetric soil water content. They have found ground water flow an important determinant of soil moisture which is important in determining hydrophytic vegetation. The hydrophytic vegetation may be found considerable distances upslope, even in well-drained soils. However, their research provides evidence that the delineation of boundaries of some wetlands is neither precise nor defensible.

In a similar vein, in Chapter 13 McMullen and Meacham compare wetland boundaries delineated in the field with the boundaries currently provided in wetland maps. The authors have reported that the maps they used were not accurate, with some maps underestimating and some overestimating the extent of wetlands. Overall, there may be more wetlands in the area compared to what is indicated in these maps. To complicate matters further, the boundaries of wetlands change, as Botts and Donn indicate in Chapter 12, where they consider the spatial and temporal dynamics of a wetland on the south shore of Lake Erie. As their findings attest, when considering coastal wetlands, geological and hydrologic processes associated with erosion and deposition can create, alter, or destroy wetlands in short periods of time.

BUFFERS

The viability of a wetland is shown to be dependent upon the maintenance of a vegetative strip. Buffer strips can be effective at reducing sediment loadings from runoff originating from upland areas. However, the width of buffer zones and the reasons for changing the widths in different jurisdictions are unclear. The effectiveness of particulate removal processes varies according to the major pathways of pollutant transport. For example, as demonstrated in Chapter 17 by McKague *et al.*, the effectiveness of alternate buffer widths in protecting a wetland downslope from an adjacent proposed urban development site is most effective when the runoff enters as a uniform sheet flow, thus maximizing the buffer strip efficiency. Norman (Chapter 18) also examines justification for different buffer widths. He reviews studies which tested the effectiveness of different buffer strip widths to control erosion and sedimentation. Grass filtration has been shown to be an effective and economic first stage procedure for reducing sediment in floodwaters. However, in the final analysis the removal efficiency is a function of width, slope, and soil permeability.

Uusi-Kämppä and Yläranta in Chapter 15 examine the effect of buffer strips in controlling soil erosion. They compare losses of nutrients and total solids from planted areas with and without buffer strips. They indicate that dense vegetation can be effective in minimizing the pollution of watercourses. The mowing and removal of grass late in the growing period decreased losses of total solids and phosphorus.

Phillips raises an interesting array of concerns in Chapter 14 by examining the question of wetlands as buffers and buffers for wetlands. The effectiveness of wetlands for attenuating water quality is improved by delayed passage through the buffers. Phillips discusses a riparian buffer delineation equation and a wetness index as a means to characterize the effectiveness of buffers.

In Chapter 10 Cox presents an interesting discussion of wetlands in riparian areas. She prefers to use the term riparian "area" over the more conventional term of riparian "zone", which she notes will distinguish the concept of riparian wetlands as an artificial unit of the landscape conceived for management or planning purposes from other ecological patches of the landscape. She points out the need for management goals as determinants of functional boundaries of wetlands.

Finally, Chapter 16 by Fraser *et al.* re-emphasizes the points made by Carter and Warner that wetlands are three-dimensional landforms. The subsurface water regime, though out of sight, cannot be ignored in the context of planning for aggregate extraction sites. Depending on the specific hydrological setting of wetlands, alterations to the natural ground water flow patterns as a consequence of changes caused by aggregate extraction activities can have serious impacts on wetlands that may be considerable distances away from the extraction site. Hence, buffer zones established on the land surface must take into account processes occurring at depths in the subsurface if wetlands are to be protected.

Clearly, this symposium represented an important step in terms of identifying the current status of our understanding of environmental gradients, boundaries, and buffers. However, it is equally apparent that much remains to be done. There is considerable need to accelerate wetland scientific research which can provide the knowledge and understanding of wetland functions (Wedeles *et al.*, 1992). Further, this symposium has provided an opportunity for the participants to see linkages between science, public policy, and professional practice.

REFERENCES

Canny, M., 1981. A universe comes into being when a space is severed: Some properties of boundaries in open systems. In *Proceedings of the Ecological Society of Australia* 11:1–11. Edited by M. J. Littlejohn and P. Y. Ladiges.

Greeson, P. E., Clark, J. R., and Clark, J. E., 1979. Wetland values and functions: American Water Resources Association. Minneapolis, Minnesota.

Krieger, K. A., Klarer, D. M., Heath, R. T., and Herdendorf, C. E., 1992. Coastal wetlands of the Laurentian Great Lakes: Current knowledge and research needs. *Journal of Great Lakes Research* 18:525–528.

Lemay, M. and Mulamoottil, G., 1984. A study of changing land uses in and around Toronto waterfront marshes. *Urban Ecology* 8:313–328.

Maltby, E., 1986. Waterlogged wealth. International Institute for Environment and Development, London.

Podniesinski, G., Deitz, K., Fisher, A., and O'Reilly, J., 1995. Wetland restoration of abandoned agricultural wetlands. *Great Lakes Wetlands* 6:1–6.

Ramsar Newsletter, 1994. The newsletter on the Convention on wetlands. 19, 1.

Richardson, C., 1985. Pocosins: Vanishing wastelands for valuable wetlands. *Bioscience* 33:626–633.

Rubec, C., 1994. The evolution of wetland policy in Canada. In *Wetland Policy Implementation in Canada*. Proceedings of a National Workshop, Stonewall, Manitoba. Report 94-1.

Snell, E., 1987. Wetland distribution and conversion in southern Ontario. Inland Water and Lands Directorate, Environment Canada. Working Paper 48.

Stadel, A., Evans, M., Kuiper, L., Mojgani, S., Weber, S., White, A., and Mulamoottil, G., 1995. Wetlands on trial: Ontario Municipal Board decisions affecting wetlands 1980–1993. *Canadian Journal of Public Administration* 38:222–241.

Wedeles, C. H. R., Meisner, J. D., and Rose, M. J., 1992. Wetland science research needs in Canada. Wetlands Research Centre, University of Waterloo, Waterloo, Ontario.

Williams, M., 1990. *Wetlands: A Threatened Landscape*. Blackwell, Oxford.

ENVIRONMENTAL GRADIENTS, BOUNDARIES, AND BUFFERS: AN OVERVIEW

2

V. Carter

Among the greatest challenges for wetland scientists today is to supply information that can be used to improve wetland protection and management. This challenge includes designing and conducting field and laboratory experiments to supply the data needed for sound planning and management decisions, as well as interpreting and disseminating information that can be used as the basis for development of wetland policy. Protection and management of our natural resources and restoration of those resources that have been destroyed or damaged requires consideration of all three themes of environmental gradients, boundaries, and buffers.

Environmental gradients are the primary reason for consideration of buffers and boundaries — any watershed contains elevation and moisture gradients as we move from the drainage divide to the lowest elevation and from the small headwater tributaries to the larger-order mainstem. In some cases, wetlands form an ecotone between upland and deepwater habitat; in many cases, wetlands are the endpoint of ground and surface water flows, and sometimes the wetland itself is large enough to contain significant environmental gradients. Materials and energy flow along these environmental gradients across the boundaries and buffers.

The overview begins with an emphasis on the importance of wetland hydrology and the need for further research, especially in ground water hydrology and ground water/surface water interactions. Hydrology, of course, is the major driving force behind wetland formation and maintenance — hydrophytic vegetation and hydric soils exist because of the persistence and availability of water on the land surface. Wetlands do not exist without water, and this has been shown in situations where water has been diverted, inadvertently or purposefully, for other uses. Many discussions of spatial patterns portray landscape elements or patches as two-dimensional areas pieced together to form a landscape, but this view is not

1-56670-147-3/96/$0.00+$.50
© 1996 by CRC Press, Inc.

particularly appropriate for wetlands. It is important to think in three dimensions when considering wetland hydrology. Johnston and Naiman (1987) illustrate the three-dimensionality of the boundaries involved in the consideration of material and energy flows in ecosystems. Some of these boundaries are lateral — for example, the boundary between a lake and a wetland — whereas others are surficial or vertical — for example, the boundary between the water and benthic sediment or the boundary between the atmosphere and the surface of the wetland. One reason for thinking three-dimensionally is the importance of ground water in forming and maintaining wetlands and in influencing wetland ecosystem dynamics. Although wetland science has come a long way from our initial attempts to understand wetland hydrology, only a few good hydrologic studies of wetlands have been published. The presence and influence of surface water is fairly obvious to many scientists and managers: we can conceptualize channeled and sheet flow and we have a basic understanding of the governing controls. Ground water is much more difficult to understand — we are saddled with the old notion that water will only flow downhill, whereas in reality water flows from areas of higher head to areas of lower head, and water can flow downward, as in recharge areas, upward as in discharge areas, and laterally as it moves through aquifers.

Riparian zones and riparian buffers provide good examples of landscape situations for three-dimensional thinking specific to wetlands and ground water. In some studies of riparian wetlands, the basic assumption or primary hypothesis is that water flows laterally from an upland — for example, a farm field, then through the riparian zone to the stream. As it moves through the riparian zone, nitrate and perhaps other constituents are removed before they enter the stream. Recent work by the U.S. Geological Survey (USGS) has shown that this may not always be the case.

Bohlke *et al.* (1991) and Bohlke and Denver (written communication, 1994) compared the history and fate of nitrate in ground water in two small adjacent watersheds on the Atlantic Coastal Plain of eastern Maryland by the combined use of chronologic indicators (tritium, chlorofluorocarbons), chemical indicators (major ions, dissolved gases), and isotopic indicators (^{15}N, ^{13}C, and ^{34}S). The soils in this area are deep and sandy and there is a riparian forested wetland along the edge of each stream. Fertilizers applied on agricultural fields in the recharge areas move downward through the permeable soils and eventually end up in the streams. This study showed that discharge into both streams showed ground water that was recharged in the fields from pre-1940 to post-1975; the average ground water residence time is 10 to 20 years. Ground water remained oxic and retained nitrate throughout most of the surficial aquifer and discharged relatively unaltered into one of the streams, which had a relatively high nitrate concentration (9 to 10 mg/L as nitrogen). In contrast, the waters were largely reduced and denitrified within calcareous glauconitic marine sediments near the bottom of the surficial aquifer before discharging into the second stream, which had a relatively low nitrate concentration (2 to 3 mg/L as nitrogen). In both cases, most of the nutrient-laden water from the fields flowed under the shallow riparian zone and discharged upward into the stream bottom (Bohlke *et al.*, 1992; Bohlke and Denver, written communication, 1994).

Bohlke and Denver (written communication, 1994) and Hamilton and Shedlock (1992) concluded that riparian forests and wetlands and organic-rich soils in stream valleys may not react with nitrate-bearing water that follows relatively deep flow paths — this water bypasses the riparian zones before discharging into the streams. Precipitation provides much of the nutrient loading to these riparian zones, and that appears to be one reason for nitrate concentrations in the ground water beneath the riparian zones to be low. In a watershed with shallow permeable soils underlain by bedrock or much less permeable soils, the water may flow laterally through the riparian zone and be affected by microbial processes and plant uptake, but this is not the case in all watersheds.

Multidisciplinary and multicomponent approaches are needed to resolve the relative importance of land use history and water/aquifer reactions on the distribution of nitrate in ground water. Wetland scientists need to identify the sources of ground water and quantify the inputs and outputs in order to estimate nutrient budgets. Riparian buffers are important for preventing erosion, controlling movement of surface water and sediment into wetlands, and improving the quality of surface water, but their role in improving the quality of ground water is less clear.

Wetland boundaries and their precise location have always been problematic. Johnston et al. (1992) have explained that although locating a boundary may be relatively simple where a significant discontinuity exists in a landscape gradient, boundary location is far more difficult and arbitrary where the environmental gradient is gradual. In the U.S. wetlands are being mapped and inventoried by the Fish and Wildlife Service's (FWS) National Wetland Inventory, using the classification of Cowardin et al. (1979). The regulation of wetlands, however, is the responsibility of the U.S. Environmental Protection Agency, the U.S. Army Corps of Engineers (COE), and the U.S. Soil Conservation Service (SCS), with the assistance of the FWS.

In the U.S. today there is considerable controversy about the delineation of wetlands for jurisdictional purposes. Three versions of a delineation manual have been published in an effort to make wetland boundary placement an objective repeatable procedure — the 1987 manual (U.S. Army Corps of Engineers, 1987), the 1989 manual (Federal Interagency Committee for Wetland Delineation, 1989), and a proposed 1991 manual published in the *Federal Register*, but never used for delineation purposes. Delineation criteria are established primarily for vegetated wetlands and delineations are based on a three-parameter approach requiring the presence of hydric soils, hydrophytic vegetation, and wetland hydrology, except in specified cases. A list of soils considered hydric and a list of hydrophytes grouped according to water tolerance are available (Reed, 1988). The hydrophytes are grouped into four categories related to water tolerance; the remaining plants are considered upland plants (Table 1). Some wetlands, for example the red maple swamps of the northeastern U.S., are dominated by facultative hydrophytes — those are plants that can grow equally well in wetlands and uplands. Wetland hydrology is the most difficult component to define and detailed information on the duration and frequency of inundation or soil saturation is rarely available.

Because of the current controversy, the 1987 manual is currently in use for wetland delineation in the U.S. To meet the criterion for wetland hydrology, it

Table 1   National Wetland Plant List Categories Based on
Frequency Occurrence in Wetlands and Weights
Associated with Each Category

Category	Frequency of occurrence in wetlands (percent)	Weights
Obligate Hydrophyte (OBL)	>99	1
Facultative, Wetland (FACW)	67–99	2
Facultative (FAC)	34–66	3
Facultative Upland (FACU)	1–33	4
Upland (UPL)	<1	5

requires that an area be periodically inundated or that its soils be saturated during the growing season. Evidence of these conditions can include secondary indicators of flooding (drift lines, silt lines on trees, blackened leaves) and any available hydrologic records used with best professional judgment. Soil saturation is measured by presence of water within the root zone, generally the top 30 cm. The 1989 delineation manual requires only seven consecutive days of flooding or saturation at or near the surface and also makes use of secondary indicators of flooding, available records, and best professional judgment. The proposed 1991 manual would require 15 consecutive days of flooding during the growing season, with required documentation of this flooding through existing records such as stage gauging. Further, to meet the saturation criterion, it would require water standing to the surface in an unlined borehole for several days. The scientific literature generally does not strongly support any of these numerical requirements — few studies have been done on the exact duration of flooding necessary to form hydric soils or result in hydrophytic vegetation. The concept of requiring a period of consecutive flooding of a specified length does not take into account the possibility of recurrent short-term flooding and its effects on both soils and vegetation. Unless water is found in a pit or well dug into the root zone, there are almost no secondary indicators of saturation in the root zone except the presence of oxidized rhyzospheres in soils.

Boundary determination is not straightforward when gradients are gradual and vegetation is tolerant of both wet and dry conditions. The Great Dismal Swamp is an 84,000-ha forested wetland on the Virginia-North Carolina border. It is bounded on the west by an ancient beach ridge, the Suffolk Scarp, and the gradient toward the east from the bottom of this scarp is slightly less than 19 cm/ km. The U.S. Geological Survey (USGS) has studied the western boundary or transition zone of the swamp in cooperation with the Fish and Wildlife Service (FWS), the Corps of Engineers (COE), and the Soil Conservation Service (SCS). Data collected on four transects ranging from 400 to 625 m long across the western transition zone have been used for a variety of boundary determinations (Carter et al., 1988; 1994). Data were collected on the hydrology, soils, and vegetation along these transects to determine the boundaries which could be established by using these components individually and in combination. The location of the wetland boundary along the Suffolk Scarp was not obvious. Some observers placed it at the edge of the wettest part of the swamp that could be seen on aerial photos and was actually near the base of the scarp. Others have located the boundary at the edge of the agricultural fields at the top of the scarp.

In the first study (Carter *et al.*, 1988), the transects were divided into 25-m increments for analysis and the vegetation data were analyzed by using the weighted averages approach developed by Wentworth *et al.* (1988). This approach makes use of the plant list for wetlands developed by Reed (1988) to supplement the Cowardin *et al.* (1979) classification system. Numerical values from 1 to 5 are assigned to the hydrophyte groups and a weighted average for the increment is calculated from the cover values for all species found. A weighted average <3.0 indicates wetland and a value ≥3.0 indicates upland. Wentworth *et al.* (1988) suggest that numbers between 2.5 and 3.5 require the supplemental use of hydric soils and hydrology indicators as verification of wetland status. We chose to call increments with weighted averages ≤2.5 wetland, between 2.5 and 3.0 transitional, and ≥3.0 upland. The transition zone of the Great Dismal Swamp is dominated by facultative species: black gum, red maple, and sweet gum. A weighted average for each increment was calculated by (1) using all the species found in the increment; (2) using just trees, just shrubs, and just understory vegetation; (3) removing the ubiquitous species; or (4) using local ecological index values based on our opinion of the correct hydrophytic status of the plants. When all species were combined most of the increments along the transects were designated wetland or transitional with very few increments, sometimes at the upland end of the transects and sometimes isolated in the middle of the transects designated upland. The other treatments gave different results. The general conclusion was that the method of weighted averages did not provide for an objective placement of an absolute boundary but served to focus attention on the transition zone, where supplementary information on soils and hydrology data were needed.

In a recent study, Carter *et al.* (1994) hypothesized that the western edge of the Great Dismal Swamp, across which the transects were sampled, is composed of three zones: the wetland itself; the ecotone or transition zone which is still wetland, but has vegetative characteristics similar to both upland and wetland; and upland. Ordination techniques were used on the vegetation data to separate clusters of increments similar in composition — vegetation boundaries were based upon these clusters. Soil boundaries were established at half the distance between soil pits that contained hydric soils and those that contained nonhydric soils, and hydrologic boundaries were placed at elevations halfway between well nests where water was in the root zone ≤25% of the growing season and those where water was in the root zone >25% of the growing season. This cutoff point for the wetland boundary is difficult to compare with that of the 1987 or 1989 delineation manual because it actually considered the duration of saturation in the root zone as measured in water-table wells. The measurements showed that ground water was discharging near the base of the scarp and maintaining moisture in the peat and mineral soils — this ground water infiltrated in a recharge area west of the swamp and moved through basal sand and gravel deposits that were laid down during past sea-level highs, and overlay relatively impermeable deep clays. Water levels were within the root zone (top 30 cm) for 25 to 100% of the growing season in the wetland and the ecotone and <25 to 50% of the growing season in the upland sites as classified by vegetation. We concluded that soils and vegetation were sufficient to delineate vegetated wetlands unless the hydrology of the site has been substantially altered, and suggested that in cases where the vegetation is primarily facultative or the hydrology has been altered, it might be more appropriate to establish a boundary zone or buffer zone rather than a single line.

Changes in vegetation along an environmental gradient are influenced strongly by disturbance (mostly logging in the area of Great Dismal Swamp). The weighted averages method for vegetative determination of wetland status that is used more widely than ordination gives an uncertain area between 2.5 and 3.5 as a result of a preponderance of facultative vegetation or a mixture of upland and wetland species. In this area, information about soils and hydrology is necessary to classify areas and make boundary decisions. This means that regulatory decisions in areas where facultative species are dominant rely heavily on soils, which of course hold the historic record of inundation and do not necessarily reflect recent water regimes.

Hydrologic information of the type gathered in this study is not generally available for most wetlands. Buffer strips around some of our larger wetlands based on an analysis of soils and vegetation might prove more realistic than sharp linear boundaries, and would encompass at least part of the area where vegetation is primarily facultative or the hydric status of soils is uncertain.

The USGS has just completed a four-year study of the floodplain wetlands along four rivers in northern Florida (Light *et al.*, 1993). A study of the Apalachicola floodplain wetlands was completed earlier (Leitman *et al.*, 1984). In the recent study, soils, vegetation, and hydrology were analyzed and state and federal wetland determinations were evaluated. The topographic relief and the duration of flooding varied from site to site, but the plots were placed in both low areas such as depressions, sloughs, low plains or terraces, and higher areas such as levees or high terraces at all sites. The low terraces, sloughs, and depressions all were classified as wetlands according to both the 1989 federal delineation manual and Florida delineation criteria, primarily on the basis of soils and vegetation, without the benefit of hydrologic records and known elevations which were not generally available. The hydrologic information available from the study confirmed these classifications. On the basis of soils, vegetation, and secondary indicators of hydrology, the high terraces on the Ochlockonee River were not determined to be jurisdictional wetlands by either the state or the federal criteria and the high terraces in the Aucilla River were not jurisdictional wetlands for the state. When direct hydrologic evidence (stage records) was consulted, these jurisdictional determinations were reversed and both the high and low terraces met wetland criteria. If the 1987 federal manual or the 1991 proposed federal delineation manual had been used none of the high terraces would have met the criteria for wetlands. The Telogia River low plain, dominated by tupelo and flooded six to ten times per year, only marginally met 1989 delineation manual criteria. It would not meet 1987 or proposed 1991 criteria because the duration of the longest flood event was typically less than one week and because sandy surface layers dry out rapidly, leaving the mucky layers in the root zone saturated for long periods, but not to the surface.

About 25 to 30% of the Apalachicola River floodplain (14,000 ha) is dominated by sweetgum, sugarberry, and water oak located on natural levees and on higher flats and ridges in the interior of the floodplain (Leitman *et al.*, 1984). It is inundated about 8% of the year or a total of 29 nonconsecutive days. Because the longest consecutive flooding is 7 to 14 days, it would not meet the criteria for wetland according to the 1987 or proposed 1991 manuals.

Studies of the Ochlockonee floodplain as a fish habitat (Leitman *et al.,* 1991) showed that 31 fish species, 75% of the total channel species, were found on the floodplain. In addition, seven common floodplain species were rarely or never found in the main channel and apparently depended on floodplain habitats for survival. The habitat requirement for fish probably includes the annual, low-duration flooding of the upper terraces. Fish tend to be opportunists, not only spatially but also temporally, fasting when food is scarce and feasting during floods when food supplies are greatly increased. Therefore, managers considering the living resources of entire wetland ecosystems may not be able to apply the strict numerical criteria in the delineation manuals for resource protection.

The South Florida wetland ecosystem is a national wetland resource extending from the Kissimmee River (north of Lake Okeechobee) south to Florida Bay. The gradient south of Lake Okeechobee is about 9.8 cm/km. From north to south the hydrology of the South Florida ecosystem has been dramatically affected by human activity and changes in hydrology have affected the entire ecosystem. The Kissimmee River was channelized in the early 1900s, resulting in loss of wetlands and increased flooding in Lake Okeechobee. Because of increased flooding Lake Okeechobee has been ringed with dikes that prevent a southward flow of water into the sugar-cane fields and toward the Everglades. The vegetable and sugar-cane fields south of Lake Okeechobee are draining nutrients into the southern part of the ecosystem and the sawgrass prairie is slowly being replaced by cattail. The diking, ditching, and draining to divert and control water and encourage agriculture have interrupted the flow of water to the Everglades and changed the seasonal patterns of flooding (Craighead, 1971). The Everglades have become drier and bird life is less abundant. In Florida Bay water clarity has been reduced, loads of nutrients and suspended sediments appear to be greater, much of the seagrass is dying, and the ecology of the bay is being increasingly dominated by phytoplankton (Boesch *et al.*, 1993). The health of the reefs adjacent to the Florida Keys is a matter of increasing concern. The restoration of this wetland ecosystem will challenge our ability to consider whole ecosystems, to think three-dimensionally, to establish linkages, to think seriously about boundaries and buffers, and to conduct relevant interdisciplinary research.

The federal government is moving to restore this extensive ecosystem. Several agencies, including the South Florida Water Management District, are beginning a multimillion dollar effort to restore the Everglades. This problem is so big that the U.S. government has recognized that restoring wetland ecosystems requires interdisciplinary and multidisciplinary research — partnerships that reach across agencies and into the university community, and relevance in research, which is research applied to solving some of the tremendous problems we have created.

In conclusion, soils, vegetation, and hydrology are the major wetland components we have to consider to address the material and energy fluxes across wetland boundaries and along wetland gradients. Wetland hydrology is a fruitful arena for more research as we need to know more about sources of water, ground water/surface water interactions, and seasonal and yearly fluctuations in water supply. Ground water, surface water, and geology are important considerations that force us to think of landscapes in three dimensions. Ecosystem dynamics and the linkages between and within wetlands must be considered for wetland restoration. Opportunities for interdisciplinary research involving gradients, boundaries, and buffers will become available. Partnerships and relevance are the new approaches to ecosystem research in the U.S. government agencies.

REFERENCES

Boesch, D. F., Armstrong, N. E., D'Elia, C. F., Maynard, N. G., Paerl, H. W., and Williams, S. L., 1993. Deterioration of the Florida Bay Ecosystem: An Evaluation of the Scientific Evidence, South Florida Water Management District, 10 p.

Bohlke, J. K., Denver, J. M., Phillips, P. J., Gwinn, C. J., Plummer, L. N., Busenberg, E., and Dunkle, S. A., 1991. Combined use of nitrogen isotopes and ground-water dating to document nitrate fluxes and transformations in small agricultural watersheds, Delmarva Peninsula, Maryland (abstract). EOS, *Transactions of the American Geophysical Union* 73:140.

Carter, V., Gammon, P. T., and Garrett, M. K., 1994. Ecotone dynamics and boundary determination in the Great Dismal Swamp, *Ecological Applications* 4:189–201.

Carter, V., Garrett, M. K., and Gammon, P. T., 1988. Wetland boundary determination in the Great Dismal Swamp using weighted averages. *Water Resources Bulletin* 24:297–306.

Cowardin, L. M., Carter, V., Golet, F. C., and Laroe, T., 1979. Classification of wetlands and deepwater habitats of the United States. *U.S. Fish and Wildlife Service* FWS/OBS-79/31, 131 p.

Craighead, F. C., 1971. The Trees of South Florida: Volume I. The Natural Environments and Their Succession. University of Miami Press, Coral Gables, Florida, 212 p.

Federal Interagency Committee for Wetland Delineation, 1989. *Federal Manual for Identifying and Delineating Jurisdictional Wetlands*, U.S. Army Corps of Engineers, U.S. Environmental Protection Agency, U.S. Fish and Wildlife Service, and U.S.D.A. Soil Conservation Service, Washington, D.C., Cooperative Technical Publication, 76 p.

Hamilton, P. A. and Shedlock, R. J., 1992. Are fertilizers and pesticides in the groundwater? *U.S. Geological Survey Circular* 1080, 15 p.

Johnston, C. A. and Naiman, R. J., 1987. Boundary dynamics at the aquatic-terrestrial interface: the influence of beaver and geomorphology. *Landscape Ecology* 1:47–57.

Johnston, C. A., Pastor, J., and Pinay, G., 1992. Quantitative methods for studying landscape boundaries, in Hansen, A. J. and di Castri, F., eds., *Landscape Boundaries*, Springer-Verlag, New York, pp. 107–125.

Leitman, H. M., Darst, M. R., and Nordhaus, J. J., 1991. Fishes in the forested floodplain of the Ochlockonee River, Florida, during flood and drought conditions, *U.S. Geological Survey Water-Resources Investigations Report* 90-4202, 36 p.

Leitman, H. M., Sohm, J. E., and Franklin, M. A., 1984. Wetland hydrology and tree distribution of the Apalachicola River flood plain, Florida, *U.S. Geological Survey Water-Supply Paper 2196-A*, 52 p.

Light, H. M, Darst, M. R, MacLaughlin, M. T., and Sprecher, S. W., 1993. Hydrology, vegetation, and soils of four north Florida river flood plains with an evaluation of State and Federal wetland determination., *U.S. Geological Survey Water-Resources Investigations Report* 93-4033, 94 p.

Reed, P. B., Jr., 1988. National list of plant species that occur in wetlands, St. Petersburg, Florida. *U.S. Fish and Wildlife Service Biological Report* 88, 140 p.

U.S. Army Corps of Engineers, 1987. *Corps of Engineers Wetlands Delineation Manual*, U.S. Army Corps of Engineers Waterways Experiment Station, Wetlands Research Program Technical Report Y- 87-1, 100 p.

Wentworth, T. R., Johnson, G. P., and Koligiski, R. L., 1988, Designation of wetlands by weighted averages of vegetation data: A preliminary evaluation. *Water Resources Bulletin* 24:389–396.

3   WETLANDS AND ENVIRONMENTAL GRADIENTS

M. M. Holland

ABSTRACT

Most landscapes contain wetland ecosystems that form transitions (ecotones) between upland and open water ecosystems. In the U.S., the Clean Water Act regulates wetlands in order to maintain wetland functions (e.g., flood protection and water quality improvement). Theoretically, wetland areas that carry out these functions are delineated for regulatory jurisdiction. Ideally, the boundary would be drawn at the point where critical functions diminish rapidly as one moves from the wetter to the drier parts of the ecosystem. Because scientific data on functional capacity are difficult to obtain, structural attributes which can be examined over shorter periods of time often are used as surrogate measures. Species composition, soil type, and hydrologic indicators all have proven to be useful indicators of wetland functioning. Thus, in delineating wetlands for any purpose, it must be remembered that wetland functions are a product of all components of the wetland ecosystem (not just vascular plants), that the wetland functions year round (not just when vascular plants are actively growing), and that critical functions (such as flood protection) will occur only at irregular intervals.

In a landscape context, wetlands and wetland ecotones are important transition zones between uplands and aquatic ecosystems. They are sites where nutrient concentrations change as water flows between terrestrial and aquatic ecosystems, and are thus important buffers between uplands and open waters. Research questions are suggested in two categories: (1) issues related to planning for maintenance of wetland functions, and (2) issues specifically related to effective wetland management.

INTRODUCTION

Environmental gradients may cause stress for organisms located at either end of the gradient. Anthropogenic or natural environmental factors or both may pose a level of stress on a species which may cause it to move to another location, to exhibit signs of physiological change, or to die. Managers of natural resources in agencies throughout the world are looking particularly for indicators of anthropogenic stress. These managers share an interest in the answers to three questions: (1) What are the stresses? (2) What are good indicators of such stresses? and (3) What can natural resource managers do to minimize, change, or alleviate the anthropogenic stresses?

This chapter presents an historical overview of discussions over the last decade on the nature and role of ecotones (Holland *et al.*, 1990; Holland *et al.*, 1991; Hansen and di Castri, 1992) and reviews recent thinking on critical research needs to understand global environmental change (Lubchenco *et al.*, 1991; Huntley *et al.*, 1991; Mooney and Sala, 1993).

CONNECTICUT RIVER OXBOWS — STUDY OF A WETLAND AND RIPARIAN FOREST COMPLEX

The author's observations on environmental gradients for the last two decades are based on a system located in northeastern U.S. (Figure 1). The system includes the lower elevational stretches of the Connecticut River and its floodplain. Work here has attempted to separate natural stresses from anthropogenic stresses in several freshwater and brackish wetlands. Gradients across a Connecticut River point bar in western Massachusetts have been examined to determine the relationship between soil characteristics and seedling establishment (Zeitler, 1978) and the relationship between water level fluctuations and plant water status (Grossman, 1981).

In the northeastern U.S., several local river basin councils working with local environmental groups have been successful in protecting fluvial ecotones. Marshes and other floodplain habitats along major streams in New England (see map, Figure 1) are spatially restricted and frequently disturbed by human activities. In western Massachusetts, extensive marsh vegetation occurs in four large oxbow lakes located along an 18-km stretch of the Connecticut River (Holland and Burk, 1982; Figure 1) used intensively for agricultural purposes. One of these oxbows has been completely protected by the Massachusetts Audubon Society, working in conjunction with the Connecticut River Watershed Council. A study of the vegetation of these marshes was included within the broader studies by Holland (1974, 1977) of the phytosociology and geological development of the three older oxbows during the period 1973–1975. In the early 1980s the relative ages of these oxbows were established (Holland and Burk, 1982), the herbaceous strata of swamp forests in the three older oxbows compared (Holland and Burk, 1984), and in the late 1980s marsh zonation patterns in the oxbows were characterized (Holland and Burk, 1990).

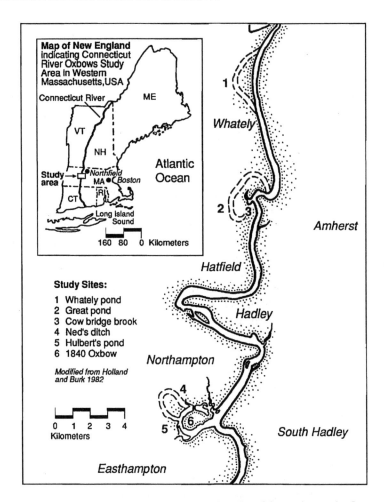

Figure 1 (A) Map of New England, indicating the location of the study area for Connecticut River oxbows in western Massachusetts (The six New England states are indicated as ME = Maine, NH = New Hampshire, VT = Vermont, MA = Massachusetts, CT = Connecticut, RI = Rhode Island); (B) Study sites are indicated as 1 = Whately Pond, 2 = Great Pond, 3 = Cow Bridge Brook, 4 = Ned's Ditch, 5 = Hulbert's Pond, and 6 = 1840 Oxbow. (Modified from Holland, M. M. and Burk, C. J., 1982.)

Holland and Burk (1990) report the results of a ten-year comparative study to examine the marshes of the three older oxbows and to compare the dominance (contribution to cover) and distribution of the vascular plant species present in 1974 with their dominance and distribution a decade later. Because of the changes documented at ponds within two of the old oxbows, particular attention was paid to possible deterioration of the marsh environments, including changes in the status of nonnative species that might be potential indicators of disturbance.

Shelford (1963) estimated that convergence to forest vegetation in oxbows and other long submergence habitats would require "not less than 1,000 years" for completion in the Mississippi floodplain, and major successional changes were not expected over the course of the decade in New England where vegetational processes proceed at a slower rate. However, Larson and Golet (1982), in an analysis of wetland habitats on the Massachusetts coastal plain, demonstrated that while deep open water and forested wetland habitats are relatively stable, marshes frequently undergo transitions to shrub swamp or forested wetlands within a period of less than 20 years. Hence, the data were also examined for any indications that successional change might be occurring, including the relative degrees of overlap between the floras of the marshes and those of adjacent oxbow swamp forest habitats. The status of uncommon or rare native plants was also noted, since oxbow habitats often support relics of formerly more widespread floras and vegetation types (Holland and Burk, 1984).

Most species of the Ned's Ditch marshes (see Figure 1) also occur in the adjacent forests, where they vary in abundance from year to year in apparent response to fluctuating water levels (Holland and Burk, 1984). The dynamic interaction of marsh and swamp forest vegetation is exemplified by the distribution within the oxbow of *Ludwigia polycarpa*, currently listed as "threatened" in Massachusetts (Sorrie, 1987). *Ludwigia polycarpa* was present in the high marsh in 1974 and became established in adjacent swamp forest quadrats by 1975. It had disappeared from high marsh by 1984 but persisted in the forest, becoming reestablished in the marsh in 1985 (Holland and Burk, 1990).

Downstream toward the mouth of the Connecticut River, changes in vegetative composition, species diversity, and community structure have been documented for two tidal brackish wetlands of the Connecticut River estuary in New England (Holland and Prach, 1993). Transect lines established in 1981 were relocated, and resampled in September 1992. Vascular macrophytes in the two marshes were surveyed; biodiversity, dominance (contribution to cover), and distribution were compared between the two marshes, and changes over the last decade were noted. Overall, percent cover increased over the course of the decade: percent cover of *Typha angustifolia* increased in both marshes, while *Phragmites communis* increased significantly only in the marsh on the eastern riverbank. The relocation of specimens of *Bidens eatonii*, a species listed by the State of Connecticut as "Special Concern", confirmed that small, healthy populations have established themselves on both sides of the river.

In the mid-1980s, many European scientists involved with either the Man and the Biosphere Programme (MAB) or the Scientific Committee on Problems of the Environment (SCOPE) were discussing the principles of landscape ecology and looking at the relationships between patches of a landscape, including movement of materials (such as energy and nutrients) from one patch to another (Risser *et al.*, 1984; Risser, 1985; Forman and Godron, 1986). As scientists sought to make generalizations about similar habitats in various locations across the globe, similar situations could be observed, for example, at the mouth of the Connecticut and the mouth of the Rhone River. Such interest in landscape ecology led to a reexamination of the word "ecotone", a word originally proposed in the early 1900s (Holland, 1988).

WETLAND ECOTONES

In a landscape context, wetlands are important (Forman and Godron, 1986; Mitsch and Gosselink, 1986). They intercept nutrients and sediment before they reach aquatic ecosystems and provide important habitats for animals, especially those that utilize aquatic and terrestrial ecosystems for feeding, protection, and reproduction. Until recently, little has been known about the ecotones between wetlands and other types of ecosystems (Johnston et al., 1992; Johnston, 1993). Here the focus is on the characteristics of wetland ecotones, especially ecological processes that occur in them. Finally, research areas requiring attention are identified in order to develop responsible management plans for wetland patches and ecotones.

Definition of Wetlands and Wetland Ecotones

Wetlands

Wetlands occur over a wide range of hydrologic conditions (Denny, 1985; Mitsch and Gosselink, 1986; Symoens, 1988) and the array of common terms used to describe them has a long history. It is necessary, therefore, to consider wetlands and wetland ecotones in the context of an accepted classification system. The system chosen here is the one adopted in the United States (Cowardin et al., 1979) and has divided the six major categories listed in that system into two types (Table 1).

Tidal wetlands — Exchanges of materials between open water ecosystems and tidal wetlands (Table 1) occur once or twice daily in response to tidal cycles. Salinity is clearly one of the most important driving factors in coastal tidal wetlands, but its importance decreases upstream as wetlands influenced by salt water give way to freshwater tidal wetlands (Odum et al., 1984).

Inland wetlands — On an areal basis, most wetlands are not along coastlines, but are found in interior regions. For example, Frayer et al. (1983) estimated that 38 million hectares, or about 95% of the total wetlands of the conterminous U.S., are inland. Exchanges of materials between the three types of inland wetlands (Table 1) and adjacent ecosystems are likely to change on a seasonal basis in response to precipitation patterns.

Table 1   Examples of Tidal and
           Inland Wetland Ecosystems

Wetland Types
Tidal Wetland Ecosystems
Tidal Salt Marshes
Tidal Freshwater Marshes
Brackish Tidal Wetlands
Inland Wetland Ecosystems
Inland Freshwater Marshes
Northern Peatlands
Swamps

Source: Holland et al. (1990).

Table 2   Comparison of the Types of Flows That Dominate Ecotones
           In Tidal and Inland Environments

Type of wetland (Patch Body)	Type of ecotone	
	Upland/Wetland	Wetland/Open Water
Tidal	Ecotone dominated by flows from wetland. Variations in flow generally small and predictable.	Ecotone dominated by flows from estuarine area. Variations in flow may be large or small but usually predictable.
Inland	Ecotone dominated by flows from upland. Large variations with high unpredictability.	Ecotone dominated by flows from open water. Large variations but mostly predictable.

Source: Holland *et al.* (1990).

Ecotones

Wetlands, like all ecosystems, have internal and external boundaries. In some instances, wetland ecotones are clearly delineated, while in others it is difficult to distinguish where one ecosystem ends and the other begins. In this paper a working definition of a MAB/SCOPE working group (Holland, 1988) is used. An ecotone is thus defined as:

> "a zone of transition between adjacent ecological systems having a set of characteristics uniquely defined by space and time scales and by the strength of the interactions between adjacent ecological systems."

The usage of "ecological system" here is analogous to that of "patch", and may refer to wetland patches and contiguous upland or open water patches. Throughout this paper the term "wetland ecotone" is analogous to "landscape boundary", "transition zone", or "wetland boundary". Although it is recognized that wetland patches are heterogeneous internally, little information is presented on ecotones within wetlands because of the lack of published papers on their characteristics.

For purposes of simplification, wetland ecotones can be characterized into the four types shown in Table 2. Most coastal and inland wetlands, referred to as patch bodies by Johnston and Naiman (1987), have lateral boundaries that connect them to adjacent upland and open water ecosystems. Johnston and Naiman (1987) use the phrase "patch body" to describe volumetric landscape units which have boundaries with upper and lower strata in addition to boundaries with adjacent patches. The connections across ecotones can be through either lateral or surficial boundaries. Surficial boundaries separate overlying patch bodies, while lateral boundaries separate patch bodies which are adjacent to each other on the same plane. Transfers across surficial boundaries have a vertical direction while transfers across lateral boundaries are primarily horizontal (Johnston and Naiman, 1987).

In both tidal and inland wetlands, vertical transfers may occur across at least four surficial boundaries (Figures 2 and 3), while horizontal transfers may occur across at least five lateral boundaries. Transfers across lateral boundaries include

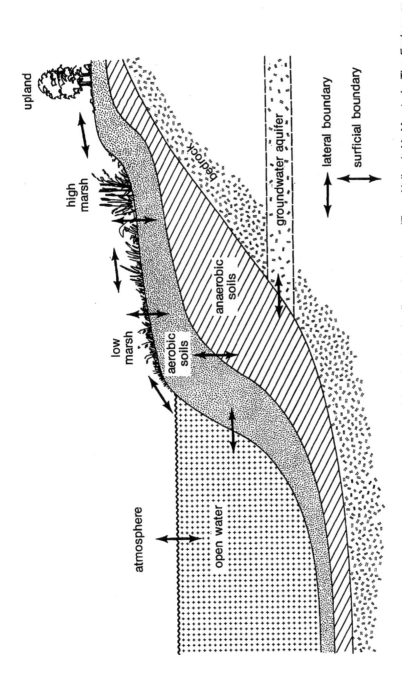

Figure 2 Generalized diagram showing ecotones between tidal wetlands and adjacent systems. (From Holland, M. M. et al., *The Ecology and Management of Aquatic-Terrestrial Ecotones*, Parthenon Publishing Group, Park Ridge, NJ, 1990. With permission.)

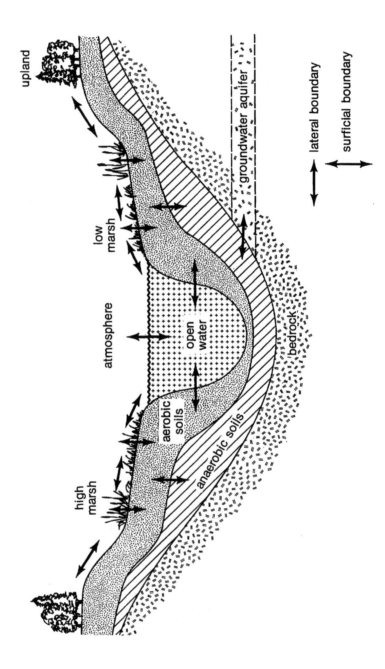

Figure 3 Generalized diagram showing ecotones between inland wetlands and adjacent systems (From Holland, M. M. et al., *The Ecology and Management of Aquatic-Terrestrial Ecotones*, Parthenon Publishing Group, Park Ridge, NJ, 1990. With permission.)

transfers from the upland to the wetland (upland/wetland ecotone), or from the wetland into open water (wetland/open water ecotone), from ground water aquifers into soils, or across vegetation zones with each zone dominated by different species (wetland/wetland ecotones). Transfers across surficial boundaries include transfers from aerobic to anaerobic soils, from aerobic soils to surficial vegetation and litter, from vegetation and litter to aerobic soils, from open water to the atmosphere, and from open water to aerobic soils (Holland *et al.*, 1990).

Wetland Patches and Ecotones as Landscape Features

The landscape has been defined as a heterogeneous land area composed of a cluster of interacting ecosystems repeated in similar form throughout (Risser *et al.*, 1984; Forman and Godron, 1986). Most landscapes contain wetland ecosystems which often form transitions (ecotones) between upland and open water ecosystems, and water flowing through the landscape will usually cross several wetland ecotones (Figure 4). Figure 4 is modified from an example in Whigham *et al.* (1988) and represents a typical U.S. coastal plain landscape. The example is given to demonstrate the fact that most landscapes contain more than one wetland and that the wetlands often form a continuum between the uplands and downstream tidal ecosystems. Figure 4 shows that there are three common types of ecotones in the landscape: upland/wetland, wetland/wetland, and wetland/open water. In the landscape shown in Figure 4, surface water and ground water would first flow from upland habitats (forests and cultivated fields) across upland/ wetland ecotones into riparian forests. The second ecotone would be the wetland/ open water ecotone between the riparian forest and the first order stream. Once in the stream channel, water would move along the wetland continuum. It would pass from riparian forests into floodplain forests and changes in water quality parameters would occur *in situ* in the stream channels except during flooding events, when water in the streams would cross the open water/wetland ecotone and would come into contact with the wetland surface (Holland *et al.*, 1990).

Water leaving the floodplain forests would, in some landscapes, pass through wetland/wetland ecotones formed where floodplain forests meet inland herbaceous wetlands. Water passing through inland herbaceous wetlands often moves as sheet flow across the wetland surface rather than in distinct stream channels. The next downstream ecotone would be the wetland/wetland ecotone at the transition between the inland herbaceous wetland and the floodplain forest. Except during flooding periods, water would remain in the streams as it moved through floodplain forests. After flowing through the wetland/wetland ecotone between the floodplain forest and the tidal freshwater wetland, interactions between the streams and the wetland would occur twice daily in response to tidal flooding. The topographically lowest ecotone in the landscape is formed between the salt marsh and the open water ecosystem. In addition to the ecotones described, all of the wetlands also have an upland/wetland ecotone at the border between the uplands and the wetlands. Overall, water moving through this typical coastal plain landscape would pass through seven different types of wetlands and

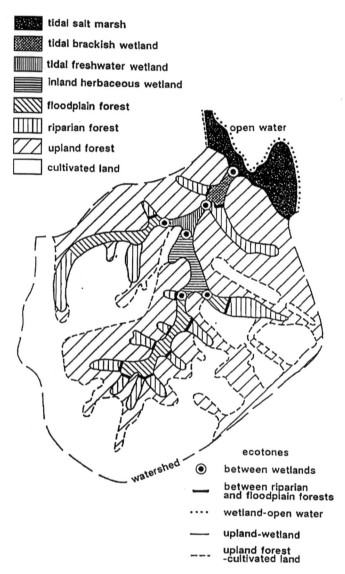

Figure 4   Generalized diagram of representative wetland patches and ecotones within a drainage basin. (From Holland, M. M. *et al., The Ecology and Management of Aquatic-Terrestrial Ecotones,* Parthenon Publishing Group, Pearl River, NY, 1990. With permission.)

across ten ecotones. As will be shown later, important changes in water quality occur as water moves across ecotones and the changes may be more dramatic in some ecotones than in others.

Upland/wetland and wetland/open water ecotones such as those shown in Figure 4 (Holland *et al.,* 1990) may be rather static, but more often they are dynamic in both space and time (Johnston and Naiman, 1987). Wetland/open water ecotones can change in space as wetlands expand into open water areas (ponds, lakes, and impoundments) or as the wetlands erode. For example, high-velocity current will often scour away sediments and vegetation, causing erosion (Johnson and Bell, 1976; Shure and Gottschalk, 1985). On the other hand, low-velocity currents will allow deposition of sediment and result in an increase in wetland vegetation (Holland and Burk, 1982, 1984).

Longer-term shifts in the position of ecotones within coastal landscapes are common and often result from variations in sea level (DeLaune et al., 1987). In climates where blanket bogs are formed, upland/wetland ecotones shift toward the uplands as wetlands expand over the landscape (Richardson, 1981). Humans can also influence the position of ecotones (Hackney and Yelverton, 1989), as can animals (Naiman et al., 1986). Changes in the position of ecotones may also occur over shorter time periods and they may be reversible (Brinson et al., 1985; Zedler and Beare, 1986).

Structural and Functional Dynamics

Physical Characteristics

Structural and functional characteristics of wetland ecotones are influenced primarily by hydrologic regimes but other factors may, at times, also be important (Johnston and Naiman, 1987). Hydrologic conditions affect many abiotic factors, including salinity, soil anaerobiosis, and nutrient availability. Since hydrology is the primary forcing function in wetlands, the biotic characteristics of wetlands and wetland boundaries are almost always directly or indirectly controlled by hydrologic changes (Niering, 1987). Differences in the magnitude (depth of submergence by tides and waves), frequency, and duration of the hydrologic interaction between ecosystems results in a variety of conditions within ecotones over differing spatial and temporal scales.

Ecotones of Tidal Wetlands

Physical features of the environment play an important role in determining the biotic characteristics of ecotones in tidal wetlands. Tides, patterns of sediment deposition and/or erosion, freshwater inputs, geological history, geographical location, and shoreline structure combined with human land use and animal activities (herbivory, burrowing, etc.) all determine the development and extent of ecotones in tidal wetlands. The interaction of tides, sediments, and geological history in determining changes in wetland ecotones has been well documented for the northeastern U.S. (Bloom, 1964; Bloom and Ellis, 1965; Redfield, 1972; Orson et al., 1987). Recent reports of accelerations in relative rates of eustatic sea level rise over the last few centuries may have significant effects on future development of coastal salt marshes (Flessa et al., 1977; Harrison and Bloom, 1977; McCaffrey, 1977; Boesch et al., 1983), and will have implications for predicting the locations of future wetland/upland ecotones.

Ecotones of Inland Wetlands

Physical features are also important in determining the development of inland wetlands and associated ecotones. Precipitation patterns, geological history, geographical location, climate, glaciation, patterns of sediment deposition and/or erosion, and runoff from surrounding uplands, combined with human land use, can all determine the development, location, and characteristics of inland wetland ecotones. The interaction of climate and geology in determining changes in wetland ecotones has been well documented for lake/sedge fen/*Sphagnum* peatland ecotones in northern regions of the Northern Hemisphere (Pigott and Pigott, 1959, 1963; Heinselman, 1970; Moore and Bellamy, 1974; Moss, 1980; Mitsch and Gosselink, 1986). Recent initiatives to understand global environmental change (Risser, 1985; Holland, 1988; Holland *et al.*, 1991; Lubchenco *et al.*, 1991; Huntley *et al.*, 1991; Hansen and di Castri, 1992; Johnston, 1993; Mooney and Sala, 1993) may recast the importance of these and similar studies in predicting future locations of upland/inland wetland boundaries should major climatic shifts occur.

Functional Characteristics

Holland *et al.* (1990) contend that ecotones are important and dynamic components of landscapes and are active sites for retention and transformation of nutrients (Peterjohn and Correll, 1984). The term retention is used to imply that materials are retained within the ecotone. Retention is usually accomplished when nutrients are assimilated into and stored in plant biomass or buried in the substrate. Transformation refers to a change in form and, in this instance, the transformation of nitrogen into NO_2 by denitrification.

Ecotones of Tidal Wetlands

In tidal wetlands (Table 2) the depth and duration of flooding is quite variable (Chapman, 1960), as are differences in nutrient characteristics and salinity patterns (Nixon and Lee, 1985; Mitsch and Gosselink, 1986). They all, however, share the common feature that the exchange of surface water between tidal wetlands and adjacent open water ecosystems is in two directions. Most water and nutrients enter and leave tidal wetlands across the surficial ecotone between open water areas (tidal creeks) and the wetland surface (Holland *et al.*, 1990). In comparison, little of the surface water exchanges vertically between the anaerobic ground water and aerobic surface water (Jordan and Correll, 1985). The vertical mixing that does occur, however, is mostly found in the lateral ecotone between the tidal wetland and tidal creek, where hydraulic conductivity is greatest (Agosta, 1985; Jordan and Correll, 1985; Yelverton and Hackney, 1986; Harvey *et al.*, 1987).

Many studies of tidal wetlands have focused on import-export characteristics (Nixon and Lee, 1985) and in some investigations ecotones (Bertness, 1984, 1985; Hopkinson and Schubauer, 1984; Mitsch and Gosselink, 1986) were considered. In general, ecotones receive high nutrient inputs, especially nutrients in particulate form, and sediment loads. Sediments near the lateral ecotone between the wetland and the tidal creek are better oxidized, and concentrations of toxic compounds are lower than those of sediments in the surficial ecotone between the tidal wetland and flooding tidal waters (Mendelssohn et al., 1982).

Exchanges of ground water with surface water occur slowly in tidal wetlands except at the ecotone between the wetland and the adjacent open water ecosystem (Agosta, 1985; Jordan and Correll, 1985; Yelverton and Hackney, 1986; Harvey et al., 1987). Between flooding events, the sediments drain more completely in this narrow zone than in high marsh areas because of differences in sediment grain size and because the area also contains numerous animal burrows. The exchange of nutrients between marsh and open water is much greater than between open water and aerobic or anaerobic soils (Jordan and Correll, 1985; Harvey et al., 1987). Most of the ground water leaving tidal wetlands comes from surface water that has infiltrated the sediments and then moved back toward the open water. The slow movement of ground water from interior areas of tidal wetlands can affect the nutrient composition of ground water that flows across the lateral boundary (Figure 2) from substrate to open water (Agosta, 1985). In a few situations, however, there is significant mixing between true ground water and open water (Valiela et al., 1978).

Exchanges across ecotones similar to those just described for saltwater tidal wetlands are probably found in freshwater and estuarine tidal wetlands, but there have been fewer transect studies conducted in those types of ecosystems. Whigham and Simpson (1978) studied a tidal freshwater wetland in New Jersey and found that there were significant changes in the nutrient composition of tidal water when it flowed over the surficial ecotone between the wetland surface and the overlying water. In another study in the same ecosystem, Simpson et al. (1981) studied patterns of distribution for heavy metals that were discharged into the wetland from a storm drain. Most metals (lead, mercury, etc.) were removed from the water column within a few meters of where the discharge pipe entered the wetland ecotone. The authors attributed the reduction in concentrations of heavy metals to deposition and sedimentation within the marginal (ecotonal) areas of the wetland.

Ecotones of Inland Wetlands

The flow of surface and subsurface water through inland wetlands is also quite variable (Cowardin et al., 1979; Mitsch and Gosselink, 1986), but unlike tidal wetlands, it is almost always in one direction (Table 2). Like tidal wetlands, the ground water is usually anaerobic (Figure 3). Aerobic conditions may be present only in a shallow surface layer. In many nontidal wetlands, periods of low hydrologic inputs are characterized by a lowering of the water table, draining of sediments, and an increase in the depth of the aerobic zone. These wetting and drying cycles play an important role in the nutrient dynamics of inland wetlands (Howard-Williams, 1985).

Processes occurring in inland wetlands are also strongly influenced by their drainage basins (Livingston and Loucks, 1979; Porter, 1981), the chemical composition of waters flowing into the basins (Richardson et al., 1978), and the sediment load of the surface water (Jaworski and Raphael, 1978; Stuckey, 1978). The amount of surface water input to most inland wetlands depends on hydrologic inputs from precipitation or from upstream ecosystems (Howard-Williams, 1985). In some inland wetlands, surface flow occurs in distinct drainage channels and there appears to be little exchange of materials across the wetland/open water ecotone between streams and adjacent wetlands except during flooding periods (Kuenzler et al., 1977; Mitsch et al., 1979; Brinson et al., 1984, 1988; Whigham et al., 1986).

Wetland/Open Water Ecotones

Verhoeven et al. (1988) studied nutrient relationships between wetlands and adjacent open water ecosystems in the Netherlands. Changes in electrical conductivity, a measure of the overall nutrient content of surface water, decreased most rapidly in the ecotone during periods when the wetland surface was flooded by water from the adjacent ditch. The authors suggested that changes were due to active plant nutrient uptake in the ecotone. Vermeer (1985), working in the same areas as Verhoeven and colleagues, found similar conductivity patterns and also showed that changes in concentrations of nutrients were greater near the ecotone between ditches and the wetland.

There have been fewer studies of nutrient transformations occurring as ground water moves across wetland/open water ecotones. Verhoeven et al. (1988) compared conductivity changes occurring in ground water moving from open water to adjacent wetlands in the Netherlands. Similar to the patterns that they found for surface water (see section above), the greatest changes occurred in the wetland ecotone for metals and nutrients in subsurface water. The ability of peat systems to retain nutrients, especially P, has been demonstrated clearly by Richardson and Marshall (1986) and it would be expected that many of the changes occur as ground water crosses wetland/open water ecotones of peat systems. Richardson and Marshall (1986) have also shown that the peat substrates become loaded with P and lose their uptake capacity. This suggests that under conditions of high nutrient loading, ecotones may have a limited ability to intercept and retain phosphorus.

Verhoeven et al. (1988) give an interesting example of how blockage of a surficial boundary between wetland vegetation and the underlying stratum can have an important impact on nutrient exchanges. In ground water discharge areas there is a net movement of water out of the aquifer, across the ecotone, and into the substratum of the wetland. Although the ground water is often polluted from nearby agricultural areas, most of the nutrients are adsorbed by the wetland substrate and do not reach the substrate surface where vegetation assimilation would occur. This situation is, however, limited to a few decades as the adsorptive ability of the peat layer will ultimately be exceeded (Richardson and Marshall, 1986).

Seischab (1987) sampled interstitial and subsurface water along a transect that crossed several vegetative ecotones ranging from deciduous and evergreen

Table 3   Percent Change in Water Quality Parameters for Pairs of Sampling Sites Within and Between Different Zones of Vegetation. Values Are Means, Independent of Whether the Changes Were Positive or Negative

Parameter	Within zone	Between vegetation zones
pH	1.68	5.25
Conductivity	5.96	37.69
Calcium	6.66	32.61
Magnesium	6.77	21.93
Nitrate	56.48	78.75

Data from Seischab (1987).

wetland forests to herbaceous and open water areas. Table 3 is a summary of the data given by Seischab (1987), where the percent change in nutrient concentrations (positive or negative) that occurred as water moved across ecotones is compared between the eight different wetland types to percent change as water moved within homogeneous patches. It is clear that the greatest variability occurred across the ecotones and demonstrates the importance of ecotones in landscapes that contain multiple wetlands.

Upland/Wetland Ecotones

Upland/wetland ecotones have been shown to be important locations where changes in water quality occur for both surface water and ground water (Peterjohn and Correll, 1984, 1986; Cooper *et al.*, 1986; Schnabel, 1986). This is particularly true when the upland areas are dominated by agricultural practices that export large amounts of sediments and nutrients. Sediments eroded from uplands primarily move in surface water, and this is the primary pathway for phosphorus movement in agricultural landscapes as most of the phosphorus is attached to sediment particles (Lowrance *et al.*, 1984; Pionke *et al.*, 1986). Nitrogen primarily moves in subsurface water as dissolved nitrate, ammonium, and organic nitrogen (Peterjohn and Correll, 1984, 1986).

The velocity of surface water slows as it passes through the surface litter layer in the ecotone, and sediments and adhering nutrients are trapped. Whigham *et al.* (1986) found that the litter layer within riparian forests retained large amounts of sediment and that some phosphorus was trapped along with the sediment. Most of the phosphorus, however, passed through the riparian zone since sediments deposited there had low phosphorus concentrations compared to the finer sediments (clays) which passed through to the adjacent aquatic ecosystem. Much of the phosphorus attached to the fine sediments was retained by wetlands further down the hydrologic gradient (Whigham *et al.*, 1988). Compared to phosphorus, most of the nitrogen is removed from subsurface water and most of the removal occurs in upland/wetland ecotones (Peterjohn and Correll, 1984). The primary mechanism for nitrogen removal in the ecotone appears to be denitrification which is driven by inputs of nitrate nitrogen in ground water from the upland, the presence of reduced sediments in the ecotone, and the high organic matter content in the soil.

RECOGNITION OF THE IMPORTANCE OF WETLAND ECOTONES

The importance of upland/wetland ecotones has been recognized by some state governments in the U.S. and legislation has been passed to insure that they are maintained (Holland and Balco, 1985). New Jersey now requires a 50-m buffer zone between uplands and wetlands (Tubman, 1988); this legislation was based in part on a scientific wetland-upland buffer delineation model (Roman and Good, 1986). It is believed that upland/wetland ecotones are most efficient in areas that are topographically lower in the landscape where soils have both aerobic and anaerobic layers, which are important requirements for denitrification (Peterjohn and Correll, 1984; Whigham et al., 1988). Ecotones that are topographically higher on the landscape might efficiently intercept sediments and nutrients in surface water but would be less efficient at intercepting nutrients in ground water because the shallow ground water would most often be aerobic.

In the highly managed landscape found in the Netherlands, the length of wetland/open water ecotone is high compared to the area of wetland. During dry periods, there is a net movement of water into wetlands from the extensive ditch systems. Polluted river water is the hydrologic source for those exchanges and wetland/open water ecotones play an important role in intercepting nutrients before they reach the interior of wetlands (Verhoeven et al., 1988). Thus, in the Netherlands, wetland/open water ecotones have an important assimilative capacity for aquatic pollutants.

KEY POINTS IDENTIFIED AS INFORMATION GAPS, STRATEGIES FOR RESEARCH, AND FOR MANAGEMENT

Identification of Gaps in Knowledge

Holland et al. (1990) present below a list of possible research questions, divided into two categories: (1) issues related to planning for maintenance of wetland functions, and (2) issues related specifically to effective wetland management.

Issues Related to Planning for Maintenance of Wetland Functions

What role does vegetation play in nutrient uptake, especially in upland/wetland ecotones during the growing season? There is evidence to demonstrate that vegetation can play an important role in removing nutrients from the substrate and ground water (Peterjohn and Correll, 1984; Fail et al., 1986). We do not know, however, how variable this characteristic is among different types of wetlands nor the importance of the ecotone itself. This last issue is of particular importance because the few data that are available suggest that upland/wetland ecotones may be important landscape features for the interception of nutrients, especially nitrogen.

How does the assimilative capacity of the wetland ecotone compare with the assimilative capacity of the wetland patch? Most investigations have focused on wetland patches, but not on ecotones. However, evidence (Holland et al., 1990) suggests that the wetland boundary may be the most important part of the wetland. Is this observation true and is it true for all types of wetlands?

Can the assimilative capacity of wetland patches and ecotones be enhanced or maintained through management? This question is obviously important yet there is little information directly addressing it.

At what temporal or spatial scale are research results most useful for decision making and management? Wetland research takes time, can be expensive, and usually only considers one wetland at a time. Most questions asked of wetlands ecologists, however, deal with issues at several scales, many of which focus on landscape issues (di Castri and Hadley, 1988; Holland, 1988). It would be productive to undertake a research project covering several hierarchical levels and considering both wetland patches and ecotones. An example would be to consider all wetlands in a drainage basin (Figure 4). The goal would be to determine which wetlands are most important in terms of intercepting nutrients and sediments and which wetland ecotones are most important. A project of this type will also have the potential to identify emergent processes as one moves from the level of a single wetland boundary to a wetland and then a series of wetlands in a drainage system.

Issues Related to Effective Wetland Management

At what level of human investment have ecotones been maintained and restored in the past and is there any evidence of the positive benefits that have resulted from those actions?

Does the nutrient retention efficiency of wetland ecotones, particularly in tidal wetlands, decrease when the edge to area ratio is altered so that there is less ecotone? If the boundaries are the most important parts of wetlands, then Holland *et al.* (1990) predict there should be some relationship between the nutrient retention capacity of an entire wetland that is related to the size (area) of the boundary or the ratio of ecotone length to patch volume (patch body).

Ecotones between uplands and wetlands appear to be especially important landscape features. How does the assimilative capacity of upland/wetland ecotones vary under different topographic and geologic settings and between different types of upland and wetland ecosystems?

Are there any types of wetland ecotones that are not important? Ecotones may not be as important as the wetlands themselves in landscapes where the flow of water is primarily in stream channels and where contact between the open water ecosystem and wetland is restricted to flooding events. Therefore, water does not come into contact with the ecotone, and when it does the uptake and/or retention capacity of the ecotone is quickly exceeded.

Along definable hydrologic gradients, are wetland ecotones topographically higher in the landscape more important in retaining and transforming nutrients than ecotones lower in the landscape?

How important are wetland/open water ecotones? In landscapes where a high percentage of the surface water volume occurs in lakes and ponds, is the ratio of wetland area to open water more important than the length of wetland ecotone?

Do wetland ecotones support high biological diversity?

Is primary production higher in wetland/upland or wetland/open water ecotones than in the wetland patches themselves?

Have wetland transition zone plant species evolved several phenotypes to allow different populations within the same species to survive not only within a wetland, but also at the wetland/open water and upland/wetland ecotones as well?

Strategies for Future Research

Oftentimes, research can provide valuable information to decision makers. For scientists, perhaps the most difficult question to answer is what type of research should be done in an area of science where the information base is so small yet the need for information is so great. We believe that field research and simulation modeling both need to be done and that the work needs to be coordinated so that the information will be of use to both scientists and resource managers. For details, refer to Summary of Final Session.

Towards a Management of Wetland Patches and Ecotones

Wise use of wetland patches and ecotones requires action on a large scale, giving consideration to all factors affecting wetlands and the drainage basins of which they are a part. A good fundamental understanding of ecosystem and hydrological processes is necessary for good management (Hollis et al., 1988). Careful synthesis and integration might ultimately result in national wetland policies which include consideration of upland/wetland and wetland/open water ecotones. For details, refer to Summary of Final Session.

SUMMARY

The following research needs have been identified previously for effective management of wetland ecotones (Brinson, 1993; Ellison and Bedford, 1995; Holland and Phelps, 1988; Holland et al. 1990; Holland and Prach, 1994):

- Examine the role of wetland ecotones in influencing biodiversity,
- Generate and test indicators of various types of stress;
- Determine ecological functions of wetland ecotones under different management regimes,
- Collect additional data on life history attributes of wetland species, and modify accordingly available models of wetland community structure,
- Determine if the assimilative capacity of landscapes and their ecotones can be enhanced or maintained through management,
- Establish reference wetlands to be used not only for the purposes of characterizing and quantifying various aspects of wetland function, but also as standards to evaluate wetland construction and restoration projects,

- Understand what is needed to restore the natural ecological integrity of wetland ecotones,
- Understand the effects of cumulative impacts (e.g., how do complex aquatic systems respond to the synergistic effects of combined disturbances?).

Current Status

Based on this overview, optimal management of wetland ecotones seems best served by:

- Management based on a long-term, whole-basin perspective,
- Coordination between state, regional, and federal agencies that have management responsibilities within the same basin, including use of consistent sampling techniques,
- Coordination of public and private land owners through local river basin councils or other nongovernmental organizations,
- Development of consensus for basin-wide goals to manage for maintenance of wetland function.

Similar complex sets of problems face wetland systems throughout the world. Achievement of sustainability often requires both minimal subsidization of managed systems so they are relatively self-sufficient, and restoration of damaged systems whose goods and services are essential to human well-being (World Commission on Environment and Development, 1989). In an urban setting, "sustainable" suggests that Earth's life support systems are maintained, and that such maintenance integrates aesthetic, economic, and ecological uses of systems (Holland and Prach, 1994).

Conclusion

Effective wetland management requires action on a broad scale, giving consideration to all factors affecting those ecotones and the drainage basins of which they are a part (Holland et al., 1990; Holland and Prach, 1994). Realization that ecotones cannot be managed in isolation from upstream inputs and the flow of benefits downstream, as well as off-site, has led to the development of legislation and policies at the local and national levels in many developed countries (Roman and Good, 1986; Holland and Phelps, 1988; Tubman, 1988). There is a need for a regional or landscape perspective in conducting research as well as in managing ecological systems.

ACKNOWLEDGMENTS

The author thanks R.A. Linthurst, G. Mulamoottil, R.P. Novitski, and B.G. Warner for comments on various drafts of the manuscript.

REFERENCES

Agosta, K., 1985. The effect of tidally induced changes in the creekbank water table on pore water chemistry. *Estuarine Coastal Shelf Science* 21:389–400.

Bertness, M. D., 1984. Ribbed mussels and the productivity of *Spartina alterniflora* in a New England salt marsh. *Ecology* 65:1794–1807.

Bertness, M. D., 1985. Fiddler crab regulation of *Spartina alterniflora* production in a New England salt marsh. *Ecology* 66:1042–1055.

Bloom, A. L., 1964. Peat accumulation and compaction in a Connecticut coastal marsh. *Journal of Sedimentary Petrology* 34:599–603.

Bloom, A. L. and Ellis, C. W., Jr., 1965. Postglacial stratigraphy and morphology of coastal Connecticut. *Connecticut Geological and Natural History Survey Guidebook 1*. Hartford, Connecticut.

Boesch, D. F., Levin, D., Nummedal, D., and Bowles, K., 1983. Subsidence in coastal Louisiana: causes, rates and effects on wetlands. *U.S. Fish and Wildlife Service* FWS/ OBS 83-26, Newton Corner, Massachusetts.

Brinson, M. M., 1988. Strategies for assessing the cumulative effects of wetland alteration on water quality. *Environmental Management* 12:655–662.

Brinson, M. M., 1993. A hydrogeomorphic classification for wetlands. U.S. Army Corps of Engineers. Waterways Experiment Station. Wetlands Research Program Technical Report WRP-DE-4, 79pp.

Brinson, M. M., Bradshaw, H. D., and Kane, E. S., 1984. Nutrient assimilative capacity of an alluvial floodplain swamp. *Journal of Applied Ecology* 21:1041–1057.

Brinson, M. M., Bradshaw, H. D., and Jones, M. N., 1985. Transitions in forested wetlands along gradients of salinity and hydroperiod. *The Journal of the Elisha Mitchell Scientific Society* 101:76–94.

Chapman, V. J., 1960. *Salt Marshes and Salt Deserts of the World*. Interscience, New York.

Cooper, J. R., Gilliam, J. W., and Jacobs, T. C., 1986. Riparian areas as a control of nonpoint pollutants, in Correll, D. L., ed. *Watershed Research Perspectives*. Smithsonian Institution Press, Washington, D.C., p. 166–192.

Cowardin, L. M., Carter, V., Golet, F. C., and LaRoe, E. T., 1979. Classification of wetlands and deepwater habitats of the United States. *U.S. Fish and Wildlife Service* FWS/OBS-79/31, Washington, D.C.

DeLaune, R. D., Patrick, W. H., and Pezeshki S. R., 1987. Foreseeable flooding and death of coastal wetland forests. *Environmental Conservation* 14:129–133.

Denny, P. (ed.), 1985. *Ecology and Management of African Wetland Vegetation*. Dr. W. Junk Publishers. Dordrecht, The Netherlands.

di Castri, F. and Hadley, M., 1988. Enhancing the Credibility of Ecology: Interacting Along and Across Hierarchical Scales. *GeoJournal* 17:5–35.

Ellison, A. M. and Bedford, B. L., 1995. Response of a wetland vascular plant community to disturbance: a simulation study. *Ecological Applications* 5(1):109–123.

Fail, J. L., Hamzah, M. N., Haines, B. L., and Todd, R. L., 1986. Above and below ground biomass, production, and element accumulation in riparian forests of an agricultural watershed, in Correll, D. L., ed., *Watershed Research Perspectives*. Smithsonian Institution Press, Washington, D.C., p. 193–224.

Flessa, K. W., Constantine, K. J., and Cushman, M. K., 1977. Sedimentation rates in a coastal marsh determined from historical records. *Chesapeake Science* 18:172–176.

Forman, R. T. T. and Godron, M., 1986. *Landscape Ecology*. John Wiley and Sons, New York.

Frayer, W. E., Monahan, T. J., Bowden, D. C., and Graybill, F. A., 1983. Status and trends of wetlands and deepwater habitat in the conterminous United States, 1950s to 1970s. Department of Forest and Wood Sciences, Colorado State University, Fort Collins.

Grossman, Y. L., 1981. The Effects of River Level Fluctuations on the Water Status of *Acer saccharinum*, A Dominant of the Floodplain Forest Community. M.S. thesis. University of Massachusetts, Amherst, 89 pp.

Hackney, C. T. and Yelverton, G. F., 1989. Effects of human activities and sea level rise on wetland ecosystems in the Cape Fear River estuary, North Carolina, U.S.A., in Whigham, D. F., Good, R. E., and Kvet, J., eds. *Wetland Case Studies*. Tasks for Vegetation Science Series. Dr. W. Junk Publishers. Dordrecht, The Netherlands.

Hansen, A. J. and di Castri, F., eds., 1992. *Landscape Boundaries: Consequences for Biotic Diversity and Ecological Flows*. Springer-Verlag, New York.

Harrison, E. Z. and Bloom, A. H., 1977. Sedimentation rates on tidal salt marshes in Connecticut. *Journal of Sedimentary Petrology* 47:1484–1490.

Harvey, J. W., Germann, P. F., and Odum, W. E., 1987. Geomorphological controls of subsurface hydrology in the creekbank zone of tidal marshes. *Estuarine and Coastal Shelf Science* 25:677–691.

Heinselman, M. L., 1970. Landscape evolution and peatland types, and the Lake Agassiz Peatlands Natural Area, Minnesota. *Ecological Monographs* 40:235–261.

Holland, M. M., 1974. The Structure and Composition of Floodplain Vegetation in Ned's Ditch. M.A. thesis, Smith College, Northampton, MA, 108 pp.

Holland, M. M., 1977. Phytosociology and Geological Development of Three Abandoned Meanders of the Connecticut River in Western Massachusetts. Ph.D. thesis, University of Massachussetts, Amherst, 281 pp.

Holland, M. M., 1988. SCOPE/MAB Technical consultations on landscape boundaries: report of a SCOPE/MAB workshop on ecotones. *Biology International, Special Issue* 17:47–106.

Holland M. M. and Balco, J., 1985. Management of fresh waters: input of scientific data into policy formulation in the United States. *Verhandlungen Internationale Vereinigung Limnologie* 22:2221–2225.

Holland, M. M. and Burk, C. J., 1982. Relative ages of western Massachusetts oxbow lakes. *Northeastern Geology* 4:23–32.

Holland, M. M. and Burk, C. J., 1984. The herb strata of three Connecticut River oxbow swamp forests. *Rhodora* 86:397–415.

Holland, M. M. and Burk, C. J., 1990. The marsh vegetation of three Connecticut River oxbows: a ten-year comparison. *Rhodora* 92 (871): 166–204.

Holland, M. M. and Phelps, J., 1988. Water resource management: changing perceptions of resource ownership in the United States. *Verhandlungen Internationle Vereinigung Limnologie* 23:1460–1464.

Holland, M. M. and Prach, R. W., 1993. Vegetation dynamics in the Connecticut River estuary: a ten year comparison. Abstract, Society of Wetland Scientists/American Society of Limnologists and Oceanographers, Edmonton, Alberta.

Holland, M. M. and Prach, R. W., 1994. Sustainability of urban wetlands, in Platt, R. H., Rowntree, R. A., and Muick, P. C., eds. *The Ecological City: Preserving and Restoring Urban Biodiversity*. University of Massachusetts Press. Amherst, MA, pp. 69–82.

Holland, M. M., Risser, P. G., and Naiman, R. J., 1991. *Ecotones: The Role of Landscape Boundaries in the Management and Restoration of Changing Environments*. Chapman and Hall. New York.

Holland, M. M., Whigham, D. F., and Gopal, B., 1990. The characteristics of wetland ecotones, in Naiman, R. J. and Decamps, H., eds. *The Ecology and Management of Aquatic-Terrestrial Ecotones*. Man and the Biosphere Book Series. The Parthenon Publishing Group, Carnforth, U.K., p. 171–198.

Hollis, G. E., Holland, M. M., Larson, J., and Maltby, E., 1988. Wise use of wetlands. *Nature and Resources* 24:2–13.

Hopkinson, C. S. and Schubauer, J. P., 1984. Static and dynamic aspects of nitrogen cycling in the salt marsh graminoid *Spartina alterniflora*. *Ecology* 65:961–969.

Howard-Williams, C., 1985. Cycling and retention of nitrogen and phosphorus in wetlands: A theoretical and applied perspective. *Freshwater Biology* 15:391–431.

Huntley, B. J., Ezcurra, E., Fuentes, E. R., Fujii, K., Grubb, P. J., Haber, W., Harger, J. R. E., Holland, M. M., Levin, S. A., Lubchenco, J., Mooney, H. A., Neronov, V., Noble, I., Pulliam, H. R., Ramakrishnan, P. S., Risser, P. G., Sala, O., Sarukhan, J., and Sombroek, W. G., 1991. A sustainable biosphere: the global imperative. *Ecology International* 20:1–14.

Jaworski, E. and Raphael, C. N., 1978. Fish, wildlife and recreational values of Michigan's coastal wetlands, wetland value study phase I. U.S. Fish and Wildlife Service Region II, Twin Cities, Minnesota.

Johnson, F. L. and Bell, D. T., 1976. Plant biomass and net primary production along a flood-frequency gradient in the streamside forest. *Castanea* 41:156–165.

Johnston, C. A., 1993. Material fluxes across wetland ecotones in northern landscapes. *Ecological Applications* 3:424–440.

Johnston, C. A. and Naiman, R. J., 1987. Boundary dynamics at the aquatic-terrestrial interface: The influence of beaver and geomorphology. *Landscape Ecology* 1:47–57.

Johnston, C. A., Pastor, J., and Pinay, G., 1992. Quantitative methods for studying landscape boundaries, in Hansen, A. J. and di Castri, F., eds. *Landscape Boundaries: Consequences for Biotic Diversity and Ecological Flows*. Springer-Verlag. New York, pp. 107–125.

Jordan, T. E. and Correll, D. L., 1985. Nutrient chemistry and hydrology of interstitial water in brackish tidal marshes of Chesapeake Bay. *Estuarine and Coastal Shelf Science* 21:45–55.

Kuenzler, E. J., Mulholland, P. J., Ruley, L. A., and Sniffen, R. P., 1977. *Water Quality in North Carolina Coastal Plain Streams and Effects of Channelization*. Report Number 13-27. Water Resources Research Institute of University of North Carolina. Raleigh.

Larson, J. S. and Golet, F. C., 1982. Models of freshwater wetland change in southeastern New England, in Gopal, B., Turner, R. E., Wetzel, R. G., and Whigham, D. F., eds. *Wetlands: Ecology and Management*. International Scientific Publishers, Jaipur, India, pp. 181–185.

Livingston, R. J. and Loucks, O. L., 1979. Productivity, trophic interactions and food-web relationships in wetlands and associated systems, in Greeson, P. E., Clark, J. R., and Clark, J. E., eds. *Wetland Functions and Values: The State of Our Understanding*. Proceedings of the National Symposium on Wetlands. American Water Resources Association, Minneapolis, Minnesota, pp. 101–119.

Lowrance, R. R., Todd, R. L., Fail, J. L., Jr., Hendrickson, O. Q., Jr., Leonard, R., and Asmussen, L., 1984. Riparian forests as nutrient filters in agricultural watersheds. *Bioscience* 34:374–377.

Lubchenco, J., Olson, A. M., Brubaker, L. B., Carpenter, S. R., Holland, M. M., Hubbell, S. P., Levin, S. A., MacMahon, J. A., Matson, P. A., Melillo, J. M., Mooney, H. A., Peterson, C. H., Pulliam, H. R., Real, L. A., Regal, P. J., and Risser, P. G., 1991. The sustainable biosphere initiative: an ecological research agenda. *Ecology* 72 (2):371–412.

McCaffrey, R. J., 1977. A Record of Accumulation of Sediments and Trace Metals in a Connecticut, U.S.A. Salt Marsh. Ph. D. Dissertation. Yale University, New Haven, Connecticut.

Mendelssohn, I. A., McKee, K. L., and Postek, M. L., 1982. Sublethal stress controlling *Spartina alterniflora* productivity, in Gopal, B., Turner, R. E., Wetzel, R. G., and Whigham, D. F., eds. *Wetlands: Ecology and Management*. International Science Publications. Jaipur, India, pp. 223–242

Mitsch, W. J. and Gosselink, J. G., 1986. *Wetlands*. Van Nostrand Reinhold Company. New York.

Mitsch, W. J., Dorge, C. L., and Wiemhoff, J. R., 1979. Ecosystem dynamics and a phosphorus budget of an alluvial cypress swamp in southern Illinois. *Ecology* 60:1116–1124.

Mooney, H. A. and Sala, O. E., 1993. Science and sustainable use. *Ecological Applications* 3:564–566.

Moore, P. D. and Bellamy, D. J., 1974. *Peatlands*. Springer-Verlag. New York.

Moss, B., 1980. *Ecology of Fresh Waters*. John Wiley and Sons. New York.

Naiman, R. J., Melillo, J. M., and Hobbie, J. E., 1986. Ecosystem alteration of boreal forest stream by beaver (*Castor canadensis*). *Ecology* 67:1254–1269.

Niering, W. A., 1987. Wetlands hydrology and vegetation dynamics. *National Wetlands Newsletter* 9:10–11.

Nixon, S. W. and Lee, V., 1985. Wetlands and water quality — a regional review of recent research in the United States on the role of fresh and saltwater wetlands as sources, sinks, and transformers of nitrogen, phosphorus, and various heavy metals. Waterways Experiment Station, U.S. Army Corps of Engineers. Vicksburg, Mississippi.

Odum, W. E., Smith, T. J., III, Hoover, J. K., and McIvor, C. C., 1984. The ecology of tidal freshwater marshes of the United States east coast: a community profile. *U.S. Fish and Wildlife Service*, FWS/OBS- 83/17, Washington, D.C.

Orson, R. A., Warren, R. S., and Niering, W. A., 1987. Development of a tidal marsh in a New England River valley. *Estuaries* 10:20–27.

Peterjohn, W. T. and Correll, D. L., 1984. Nutrient dynamics of an agricultural watershed: observations on the role of a riparian forest. *Ecology* 65:1466–1475.

Peterjohn, W. T. and Correll, D. L., 1986. The effect of riparian forest on the volume and chemical composition of base flow in an agricultural watershed, in D. L. Correll, ed. *Watershed Research Perspectives*. Smithsonian Institution Press, Washington, D.C., p. 244–262.

Pigott, C. D. and Pigott, M. E., 1963. Late-glacial and post-glacial deposits at Malham, Yorkshire. *New Phytologist* 62:317–334.

Pigott, M. E. and Pigott, C. D., 1959. Stratigraphy and pollen analysis of Malham tarn and Tarn Moss. *Field Studies* 1:1–17.

Pionke, H. B., Schnabel, R. R., Hoover, J. R., Gburek, W. J., Urban, J. B., and Rogowski, A. S., 1986. Mahantango Creek watershed — fate and transport of water and nutrients, in Correll, D. L., ed. *Watershed Research Perspectives*. Smithsonian Institution Press, Washington, D.C., pp. 108–134

Porter, B. W., 1981. The wetland edge as a community and its value to wildlife, in *Selected Proceedings of the Midwest Conference on Wetland Values and Management*. Fresh Water Society, pp. 14–15.

Redfield, A. C., 1972. Development of a New England salt marsh. *Ecological Monographs* 42:201–237.

Richardson, C. J., 1981. *Pocosin Wetlands*. Hutchinson Ross Publishing Company, Stroudsburg, Pennsylvania.

Richardson, C. J. and Marshall, P. E., 1986. Processes controlling movement, storage, and export of phosphorus in a fen peatland. *Ecological Monographs* 56:279–302.

Richardson, C. J., Tilton, J. L., Kadlec, J. A., Chamie, J. P. M., and Wentz, W. A., 1978. Nutrient dynamics of northern wetland ecosystems, in Good, R. E., Whigham, D. F., and Simpson, R. L., eds. *Freshwater Wetlands: Ecological Processes and Management Potential.* Academic Press, New York, pp. 217–242

Risser, P. G., compiler, 1985. Spatial and temporal variability of biospheric and geospheric processes: research needed to determine interactions with global environmental change. *Report of a Workshop in St. Petersburg, Florida*, October 18–November 1, 1985. Sponsored by SCOPE, INTECOL, and ICSU.

Risser, P. G., Karr, J. R., and Forman, R. T., 1984. Landscape ecology: directions and approaches. *Illinois Natural History Survey, Special Publication* Number 2, Champaign, Illinois.

Roman, C. T. and Good, R. E., 1986. Wetlands of the New Jersey Pinelands: values, functions, and impacts. Division of Pinelands Research Center for Coastal and Environmental Studies, Rutgers — the State University, New Brunswick, New Jersey.

Schnabel, R. R., 1986. Nitrate concentrations in a small stream as affected by chemical and hydrological interactions in the riparian zone, in Correll, D. L., ed. *Watershed Research Perspectives.* Smithsonian Institution Press, Washington, D.C., p. 263–282

Seischab, F. K., 1987. Succession in a *Thuja occidentalis* wetland in western New York, in Laderman, A. D., ed. *Atlantic White Cedar Wetlands.* Westview Press, Boulder, Colorado, pp. 211–214

Shelford, V. E., 1963. *The Ecology of North America.* University of Illinois Press, Urbana, Illinois, 610 pp.

Shure, D. J. and Gottschalk, M. R., 1985. Litter-fall patterns within a floodplain forest. *American Midland Naturalist* 114:98–111.

Simpson, R. L., Good, R. E., and Frasco, B. R., 1981. Dynamics of nitrogen, phosphorus, and heavy metals in Delaware River freshwater tidal wetlands. *Final Technical Completion Report.* Corvallis Environmental Research Laboratory. U.S. Environmental Protection Agency. Corvallis, Oregon.

Sorrie, B. A., 1987. Notes on the rare flora of Massachusetts. *Rhodora* 89:113–196.

Stuckey, R. L., 1978. The decline of lake plants. *Natural History* 87(7):66– 69.

Symoens, J. J., ed., 1988. *Vegetation of Inland Waters.* Kluwer Academic Publishers, Dordrecht, The Netherlands.

Tubman, L. H., 1988. New Jersey's freshwater wetlands protection act. *Journal of the Water Pollution Control Federation* 60:176–179.

Valiela, I., Teal, J. M., Volkmann, S., Shafer, D., and Carpenter, E. J., 1978. Nutrient and particulate fluxes in a salt marsh ecosystem: tidal exchanges and inputs by precipitation and groundwater. *Limnology and Oceanography* 23:798–812.

Verhoeven, J. T. A., Koerselman, W., and Beltman, B., 1988. The vegetation of fens in relation to their hydrology and nutrient dynamics, in Symoens, J. J., ed. *Vegetation in Inland Waters.* Dr. W. Junk, Publisher, Vordrecht, The Netherlands, pp. 249–282.

Vermeer, H., 1985. Effects of Nutrient Availability and Groundwater Level on Shoot Biomass and Species Composition of Mesotrophic Plant Communities. Dissertation. State University of Utrecht. Utrecht, The Netherlands.

Whigham, D. F. and Simpson, R. L., 1978. Nitrogen and phosphorus movement in a freshwater tidal wetland receiving sewage effluent, in *Coastal Zone 78. Proceedings of the Symposium on Technical, Environmental, Socioeconomic, and Regulatory Aspects of Coastal Zone Management.* American Society of Civil Engineers. Minneapolis, Minnesota, pp. 2189–2203.

Whigham, D. F., Chitterling, C., and Palmer, B., 1988. Impacts of freshwater wetlands on water quality — a landscape perspective. *Environmental Management* 12:663–671.

Whigham, D. F., Chitterling, C., Palmer, B., and O'Neill, J., 1986. Modification of runoff from upland watersheds — the influence of a diverse riparian ecosystem, in Correll, D. L., ed. *Watershed Research Perspectives*. Smithsonian Institution Press, Washington, D.C., pp. 283–304.

World Commission on Environment and Development. 1989. Sustainable development: a guide to our common future. The Global Tomorrow Coalition, Washington, D.C. 400 pp.

Yelverton, G. F. and Hackney, C. T., 1986. Flux of dissolved organic carbon and pore water through the substrate of a *Spartina alterniflora* marsh in North Carolina. *Estuarine and Coastal Shelf Science* 22:252–267.

Zedler, J. B. and Beare, P. A., 1986. Temporal variability of salt marsh vegetation: The role of low-salinity gaps and environmental stress, in D. Wolfe, ed., *Estuarine Variability*. Academic Press, New York, p. 295–306.

Zeitler, P. S., 1978. Seedling Establishment of Floodplain Species at Two Sites Along the Connecticut River in Northampton, Massachusetts. B.S. thesis. Amherst College, Amherst, Massachusetts, 122 pp.

4   VERTICAL GRADIENTS IN PEATLANDS

B. G. Warner

ABSTRACT

Peatlands are considered as three-dimensional landforms which have both horizontal and vertical gradients. The vertical gradient in peatlands from older to younger, is primarily developed as a result of the accumulation of peat through time. In this chapter, a conceptual model for understanding the physical, hydrological, chemical, and biological processes regulating peat accumulation and system dynamics is presented. The pollen and spore records contained in peatlands in southern Ontario show a major change along the vertical gradient coincident with the time of arrival of the first European settlers. It appears that European settlement has had a major impact on the peatland systems in southern Ontario by causing both gains and losses. A scheme for classifying peatlands affected by human activity is presented. Such knowledge of overall peatland conditions and the magnitude of human impact is vitally important to the management, restoration, and conservation of peatland resources.

INTRODUCTION

Peatlands have both horizontal and vertical gradients. Differences in surface water pH and vegetation along horizontal gradients have been the traditional means for classifying peatlands. Two main categories, bogs and fens, are usually recognized (Gorham and Janssens, 1992). Bogs are identified on the basis of acidic waters and the dominance of *Sphagnum* mosses, and are found above flood levels of runoff from surrounding uplands and above ground water that has percolated through mineral soil. In contrast, fens are identified on the basis of circum-neutral waters rich in dissolved minerals, notably calcium, and an abundance of brown mosses that are primarily members of the family Amblystegiaceae, and vascular plant species. In reality there is a continuous gradient between bog and fen. As many as seven categories ranging from extreme rich fen, transitional

1-56670-147-3/96/$0.00+$.50

rich fen, intermediate fen, transitional poor fen, intermediate poor fen, extreme poor fen, and bog have been identified on the bog-fen continuum (Gorham and Janssens, 1992).

However, there is much more to peatlands than what can be observed along the horizontal gradient on the surface. Peatlands also contain vertical gradients. The vertical gradient from older to younger is primarily developed as a result of the accumulation of peat through time. The work of Johnston and Naiman (1987) is noteworthy because it is one of the few ecological studies that recognizes peatlands as three-dimensional "patch bodies". In addition to transitions on the land surface along horizontal gradients, there are transitions in the subsurface along vertical gradients (see Chapter 3). For peatland patch bodies, the most important of these transitions on the vertical gradient are those between vegetation/litter and oxygenated and seasonally saturated peat layers, between seasonally saturated peat layers and permanently saturated peat layers, between permanently saturated peat layers and mineral deposits/bedrock, and between open water and saturated peat layers.

Geomorphologists consider peatlands as landforms which are composed of biological materials with distinctive morphologies. As early as the middle 1800s peatland scientists described the size, area, depth, and stratigraphy of individual peat landforms (e.g., Auer, 1927; Aario, 1932; von Bülow, 1929; Kulczynski, 1949; Overbeck, 1975). Many of their descriptions contain words such as microrelief, "Kleinformen", hummock, and hollow, which were used to distinguish small-scale surface features from the large-scale peat landforms.

To ecologists, peatlands are ecosystems, and landscape ecologists describe peatland ecosystems as "patches" (see Chapter 3). Peatlands are patches having a minimum of 40 to 50 cm of saturated organic soil and peat (Cowardin *et al.*, 1979; Canadian National Wetlands Working Group, 1988). In the landscape mosaic, transitions form along horizontal gradients between upland and peatland patches, between mineral wetland and peatland patches, and between open water and peatland patches (see Chapter 3). Peatland patches also contain transitions between the different kinds of peatland patches within it.

It is not possible to separate the peat landform from the peatland ecological system since these two together give rise to and sustain the "peat landform ecological system" (henceforth referred to as the "peatland system"). As such, peatlands can be thought of as living landforms supporting an ecological system which adds peat (biological materials) to the landform, leading to growth and development.

The vertical transfer of biological materials across subsurface transitions (surficial boundaries *sensu* Johnston and Naiman, 1987; Holland, this volume) in the process of formation and accumulation of peat is an unusual and important attribute of peatland systems compared to other patches of the landscape. As peatland systems evolve and accumulate peat, they preserve a record of their past. Biotic and abiotic components of the peat reflect changes along the vertical gradient and can be used to compile a history of the peatland system as it changes in response to internal factors and to environmental factors external to the peatland system. This extended temporal framework is most useful for establishing the sequence of events which contribute to contemporary peatland systems, forms, structures, and system processes. It also provides a baseline to gauge the present and future condition of the peatland system itself and of adjacent patches in the landscape.

This chapter (a) describes the general features of the peat landform system, the characteristics of the vertical gradient, and discusses some of the ways biotic and abiotic components of the peat reflect changes along the vertical gradient over time; (b) provides a conceptual model of peatland system processes in order to understand the causes of changes in peat accumulation and system dynamics; (c) highlights changes in the pollen and spore record along the vertical gradient as a result of landscape change brought about by European settlement in southern Ontario; and (d) illustrates the potential for studying changes along the vertical gradient as an indication of the overall peatland condition and the magnitude of human impact.

THE PEAT LANDFORM AND FEATURES OF VERTICAL GRADIENTS

As with any landform, peatlands have physical **form** varying in size, shape, depth, and total mass. They can be subdivided into two subforms. The first is a constructional form where peat is actively accumulating and the peat landform is growing. Most natural peatlands probably fall into this category. The second is a destructional form where overall volume is decreasing. In the case of the latter subform, there is less peat present now than at some point in the past for reasons that may depend on internal factors which have essentially closed down active peat accumulation, or may depend on factors external to the peat landform. Coastal peatlands eroded by the sea or lake waters, or peatlands altered by humans, would demonstrate a decrease in peatland form.

Human disturbances can give rise to both contructional and destructional peat landforms. There are generally two main types of destructional forms which relate to human activities (Overbeck, 1975; Rowell, 1988; Göttlich, 1990). The first type is degenerated peatlands where the surface remains more or less intact and the surface oxic and seasonally saturated zone (acrotelm in the case of bogs), rooting zone, and living vegetation are generally degraded. The second type is cutover peatlands where the vegetation, surface oxic and seasonally saturated zone, and part of the peat in the permanently saturated zone (catotelm in the case of bogs) have been removed and peat that was once in the permanently saturated zone is left exposed on the surface.

Peat landforms are complex. Indeed, this has led to the use of such terms as "peatland complex". Different kinds of peat landforms exist which vary in surface area depending on their position relative to adjacent mineral landforms and relative to other peat landforms. The concept of the geotope utilized by Ivanov (1981) is a convenient way of subdividing the peat landform into three scales of size. The macrotopic scale refers to the position of the peat landform (i.e., peatland patch body in ecological terms) in the context of other types of landforms, mineral

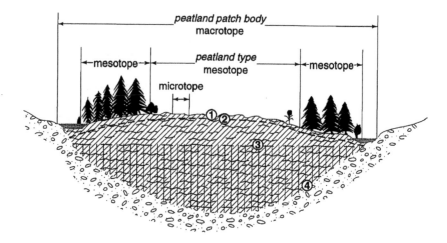

Figure 1   Schematic cross section of a raised bog illustrating the geotopic scales and possible equivalents used by landscape ecologists. Also shown are the positions of major subsurface boundaries. Transfers across these boundaries occur in both directions: 1 = represents two boundaries: transfers between living plants and litter/dead plants, and transfers between litter/dead plants and peat; 2 = transfers between the surface oxic layer and the lower anoxic layer; 3 = transfers between different peat types (e.g., *Sphagnum* to sedge peat); 4 = transfers between the peatland system and adjacent mineral deposits.

and biological, on the land surface (Figure 1). The mesotopic and microtopic scales denote units within the peatland. The mesotopic scale recognizes peatland types (*sensu* Canadian National Wetlands Working Group, 1988) with distinctive morphology, plant physiognomy (usually vascular plant strata), and water supply. The microtopic scale recognizes smaller-scale features such as hummocks, hollows, and lawns usually formed within the moss strata.

Peat landforms have various **structural** attributes. It is probably easiest to think of peat landforms as having two primary structural layers — external or surface layers which provide the materials for deeper internal layers. The external layers are important because they play an integral role in determining internal structural characteristics such as composition and arrangement of layers and the physical, chemical, and biological properties of materials within the peat landform.

The concept of the diplotelmic peatland has been suggested for raised bogs (Ingram, 1978; Ivanov, 1981). The main idea is that diplotelmic bogs are characterized by two layers — an oxic surface layer (the acrotelm) where most biological activity and water movement occurs much more rapidly than in the underlying oxygen-poor or anoxic layer (the catotelm) in which decay and water movement occur much more slowly. How applicable this is to other kinds of peat landforms remains to be explored.

Internal structures are the result of transfers of materials from external to internal layers, but deeper peatland materials are susceptible to further deformational and diagenic alterations. Since living external layers are also dependent upon the deeper internal layers for physical support, deformational changes to internal layers can also affect external layers. Transfers between these two zones represent one of the most important transitions along the vertical gradient.

PEAT ACCUMULATION AND DEVELOPMENT OF THE VERTICAL GRADIENT

The peat accumulation process is common to all peatlands and is the primary feature of the vertical gradient. The process which contributes to peatland initiation, formation, and maintenance is lucidly described by Clymo (1983, 1991) and Moore (1987). Peatlands begin to form because water is constantly at or above the mineral land surface. These conditions encourage the colonization of plants and animals specifically adapted to waterlogged soils. When these plants and animals have completed their life cycles, the remains fall at or near the level of the water during the earliest stages of peat landform development. During later stages, the remains are added to an accumulating peat mass. This process is ongoing and results in the accumulation of peat year after year. The accumulation of peat is the net result of the interactions between the quantity of dead plant biomass falling on the peatland surface, the loss of plant and peat biomass through decay, and the rate at which it sinks below the water table. Once the disarticulated plant and animal remains become saturated in the perennially waterlogged and oxygen-poor layers at depth, rapid decay is prevented and they are preserved. They do not decompose completely and remain there for long periods of time, even millennia.

The partial remains of plants and animals in the peat along the vertical peat gradient are a record back in time of biota that once lived in and around the peatland.

Cores or monoliths of peat can be cut into vertically stacked "time slices". The remains of plants and animals preserved in the slices of peat, which can range from 10×10^{-6} m (e.g., *Ambrosia* pollen) to 10^{-2}–10^{-1} m (e.g., cones, wood) in size can be identified. Almost any plant or animal species once living in the peatland can leave some evidence of its presence in the peat. The nature of past biotic communities and their hydrological and geochemical environments can be traced by examining the chemical composition of identifiable biotic remains and by substances that have been transformed by biological and geochemical processes in the soil, the bulk peat, and interstitial waters and gases. The results of these kinds of analyses are plotted along a vertical axis in diagrams which portray the past generations of biotic life.

In addition to biological components and abiotic constituents, the physical nature and state of decomposition of the peat provide important information on conditions at the time the peat was formed. Details on the biotic and abiotic parameters in peatlands can be found in Warner and Bunting (1995).

A CONCEPTUAL MODEL OF PEATLAND SYSTEM DYNAMICS

The process of peat accumulation, which occurs as the vertical transfer of materials between the external oxic layers and the internal anoxic layers, is not in static equilibrium. The process is dependent upon changes in the physical structure, chemical state, and quantity of decomposing biotic materials (mostly plant tissues). As a result of these transfers, transitions along the vertical gradient are not static but can change as peat continues to accumulate and moves mostly upward on the vertical gradient in the case of constructional peatlands or remains stationary or moves downward in the case of destructional peatlands. Unfortunately, it has been difficult to relate structural features of external and internal layers to process. Each of these two main structural zones can contain additional layers within them (Figure 2). For example, as many as five to nine sublayers have been recognized within the external layer (Clymo, 1992; van Dierendonck, 1992). Similarly, transitions within internal layers may occur between thicker units of different peat types or between much thinner units commonly referred to as black streaks.

It is necessary to understand the main processes that operate within peatland systems in order to recognize differences or interpret the causes for change along a vertical gradient. Four major groups of internal processes work together to sustain a peatland system. These are physical, hydrological, geochemical/chemical, and biological processes (Figure 3). Physical processes relate to the overall morphological form of the peatland itself and its position in the context of other landforms on the land surface. Hydrological processes are closely connected to physical processes since topography and geological setting determine the location, quantity, and origin of the water that will collect on the land surface, thus initiating peat formation and controlling its development. The water in the system will in turn have a characteristic geochemical signature. The primary difference is between waters rich in dissolved minerals vs. waters poor in minerals and is determined by whether the water has the opportunity to come in contact with mineral deposits and bedrock. Finally, the invading peatland biota which respond to the hydrology and specific geochemical signatures of the water initiate the biological processes. Biological processes are supported by the physical, hydrological, and geochemical processes. All four processes are intricately balanced and the strength of each performs in a way that supports a wide array of peatland biota that contribute to the accumulation of peat and to the great diversity of peatland systems. Together, they give rise to the peatland system which is positioned in the center of the model.

A number of factors external to the peatland system can affect some or all of the physical, hydrological, geochemical, and biological processes which will alter the peat landform accordingly. Natural geomorphic changes, i.e., sea/lake level changes, isostatic adjustments, natural fires, climate, and human activities are the most important, and there may be others. The spheres can be thought of as balloons, each of which has a certain amount of resiliency to change. The effects of external forcing factors on any one or more of the balloons will be felt by the other balloons in the system. Beyond a certain threshold the balloon(s) may shrink or expand, which may change the whole system.

Time is left "floating" on the perimeter of the model to denote the dynamism of peatland systems. Though peatland systems cannot exist without all four of these processes operating, it seems that the contribution of each may gradually change — some becoming stronger, others becoming weaker — as the peatland

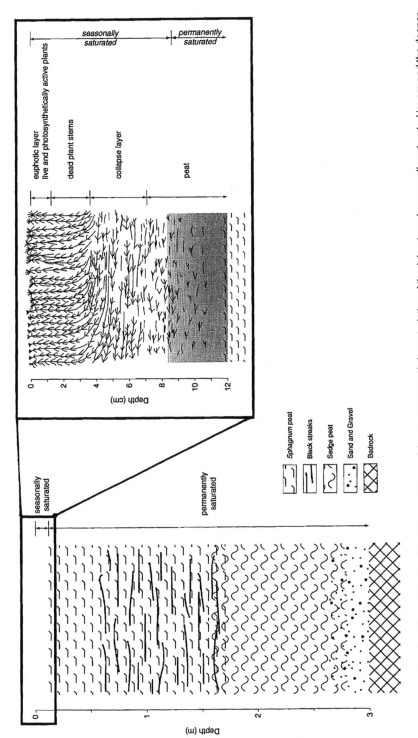

Figure 2   Sketches illustrating simplified interpretations of the structures and characteristics of the (a) external seasonally saturated layers and the deeper internal permanently saturated layers in a typical basin bog, (b) near-surface layers in a typical fen, and (c) near-surface layers in a cutover peatland having secondary peat accumulation.

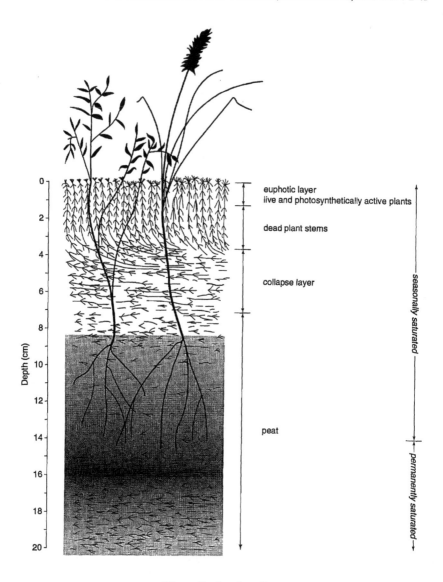

Figure 2 (continued)

ages. The rate and strength of interactions between these processes are continuous long-term processes. However, when some factor such as human activities in and around the peatland interfere, these long-term processes may intensify and speed up the rates at which they operate. The change may contribute to net gains or losses to the peatland system.

This model is useful because it illustrates the complexity of, and the interactions between, the processes sustaining the peatland system. In general, all peatlands have physical, hydrological, geochemical/chemical, and biological transitions that exist along both horizontal and vertical gradients.

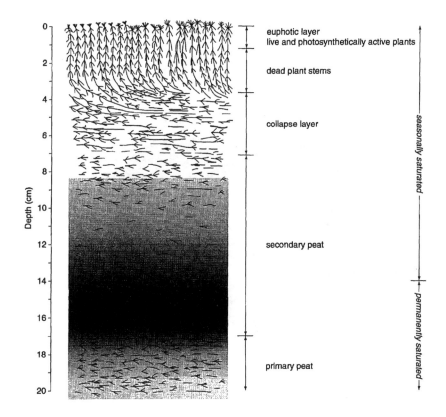

Figure 2 (continued)

CHANGES IN VERTICAL GRADIENTS AND LANDSCAPE SINCE EUROPEAN SETTLEMENT

The characteristics of the vertical peat gradients and the biotic and abiotic indicators along the gradient have been studied from various kinds of peatland systems in southern Ontario. The pollen and spores preserved in two contrasting types of peatlands reveal that one of the most notable changes along the gradient occurred at the time of arrival of European settlers. The following examples illustrate how changes in the vertical gradient can be identified and then used to speculate on the causes for the changes.

Kettle Hole Floating Peatlands

Floating peatlands, where a buoyant peat mat has developed over waterbodies in kettle holes, are a distinctive type of peatland in southern Ontario. Two major kinds of floating peatlands exist. The first type develops a buoyant peat mat along the outside edges of the basin around an area of open water in the center. A minor variation of this type is one where a buoyant mat covers the entire basin and no residual open water area is evident (Figures 4 and 5). The second type is where

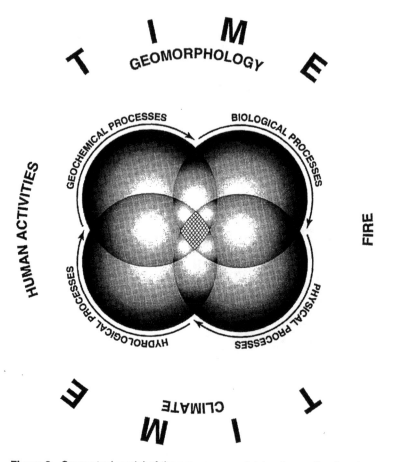

Figure 3 Conceptual model of the processes sustaining the peatland system.

a buoyant mat of peat forms an island in the center and open water encircles the edges of the basin (Damman and French, 1987; Warner, 1993).

Studies of pollen and macroscopic plant remains in peat cores taken through the existing *Sphagnum*-dominated communities of floating mats in representative examples of floating peatlands indicate an origin during the historical period (Figure 6). The abundance of *Ambrosia* pollen is a common indicator of European forest clearance and the presence of agricultural fields since *Ambrosia* is a plant which quickly colonizes sunny areas and fresh soils (see for example Figure 7, zone B). Thus, a rise in abundance of *Ambrosia* pollen, along with high proportions of pollen of weedy and herbaceous taxa of open fields and forest clearings, signifies the European arrival. Pollen of forest tree taxa is also much reduced, probably due to logging and land clearance, in comparison with the pre-European period. These analyses also show increases in spores and macroscopic remains of *Sphagnum* and other typical peatland species such as *Larix*, Ericaceae, *Ilex/Nemopanthus*, and *Menyanthes trifoliata* acompanying abundant *Ambrosia* pollen (Warner *et al.*, 1989; Warner, 1993).

Figure 4 Aerial photograph taken in 1978 of Sifton Bog, a kettle hole floating peatland in London, Ontario. A buoyant mat almost covers the entire basin.

The co-occurrence of peatland species with indicators associated with the European period show that deforestation by early European settlers caused the floating peat mats to form on the surface of what was either a lake or shallow open water wetland in the pre-European period (Warner *et al.*, 1989; Warner, 1993). Deforestation and agricultural fields likely reduced the regulating effects of forest cover on snowmelt during the spring thaw. This would have resulted in either increased surface runoff and spring floods in the undrained kettle holes or a rise in the local ground water table, or both. The peatland shrubs and mosses were favored by the rise in water levels in the kettle holes and established floating carpets of vegetation over the open water bodies.

The pattern is characteristic of floating peatlands from Delhi to Orangeville in southwestern Ontario (Warner, unpublished data). Though *Sphagnum*-dominated communities could have been present in the kettle holes before the arrival of European settlers, human alterations to the natural forest cover apparently caused changes to the local hydrological balance of kettle holes. The hydrological change triggered growth and lateral development of floating mats of peat on the surface, thereby converting open water into floating peatland communities.

The lateral spread of the floating mats across the open waterbodies in the kettle also illustrates a dramatic change along horizontal gradients during the historical period. The newly forming peatland diversified into a range of

Figure 5   A view of the floating mat vegetation of Sifton Bog. The trees are *Larix laricina* and *Picea mariana*. The shrubs in the foreground are *Chamaedaphne calyculata*. The mosses include *Sphagnum fuscum*, *S. angustifolium*, and *Polytrichum strictum*.

communities at all geotopic scales. This is one example where human impacts resulted in a gain in size and diversity in the peatland system.

Wainfleet Bog: A Degraded Peatland

True bogs are rare in southwestern Ontario. One of the finest examples is the Wainfleet Bog on the Niagara Peninsula near Port Colborne. It is Canada's southernmost raised bog and as such is nationally important owing to its latitudinal position and to the number of rare and endangered plants and animals found in it (Fraser, 1989). It is about 1030 ha in area (Figure 8), which is about one-fifth of its extent prior to the European period (Leverin, 1941; Gentilcore and Head, 1984; Nagy, 1992). The drastic reduction in size occurred mainly between 1850 and 1933 due to drainage, land conversion to agriculture, and construction of the Welland Canal. A network of drainage canals was dug in the early 1900s for peat harvesting operations in the western part of the bog. In 1941, 1093 ha of the then remaining 1416 ha of the bog was being used for harvesting peat (Leverin, 1941). The surface vegetation has been burned on several occasions; the most recent fire occurred in 1971 (Fraser, 1989).

The eastern part of the bog contains the least disturbed and most intact bog communities and has been designated an Area of Natural and Scientific Interest (ANSI). *Betula occidentalis* is the only tree species in the central part of the open bog (Figure 9). Ericaceae shrubs including *Vaccinium corymbosum, Chamaedaphne*

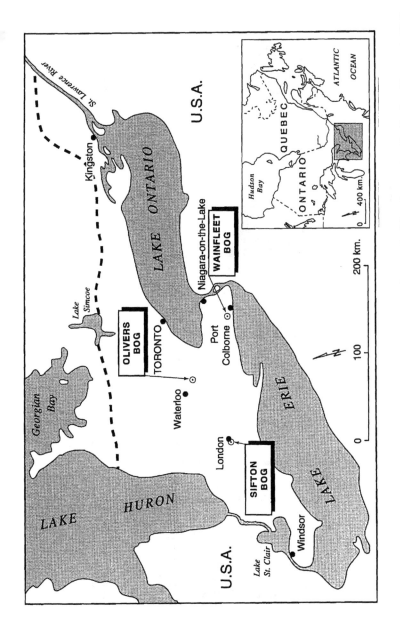

Figure 6 Map showing the location of peatland study sites referred to in the text. The dashed line demarcates the Eastern Temperate Wetland Region in the south and the Low Boreal Wetland Region in the north (according to the Canadian National Wetlands Working Group, 1988).

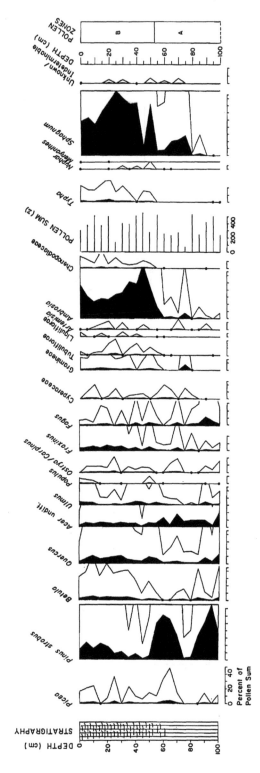

Figure 7 Abbreviated pollen diagram for Oliver's Bog near Cambridge, Ontario. This site is an example of a floating peatland with a mat developing around the periphery of the basin. (Adapted from Warner, et al. 1989).

Figure 8   Aerial photograph taken in 1987 of the Wainfleet Bog. Note the peat cuttings on the west side of the bog (right side) and the degraded east part of the bog (left side).

Figure 9   View of vegetation in the degraded bog area in the ANSI on the eastern side where the core for this study was collected (see Figure 10). The trees are *Betula occidentalis*. The shrubs are *Pyrus melanocarpa* which have invaded as a result of fire.

calyculata, Kalmia angustifolia, and *Vaccinium oxycoccus* are characteristic. A cover of *Pyrus melanocarpa* can be found in this part of the bog also. *Sphagnum fuscum, S. magellanicum, Polytrichum strictum,* and *Aulacomnium splendens* are the dominant mosses.

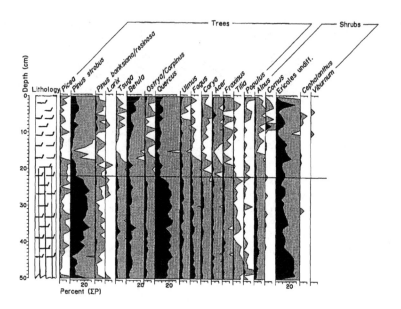

Figure 10 Abbreviated pollen diagram for Wainfleet Bog near Port Colborne, Ontario. (Adapted from Nagy, B. R., 1992).

The pollen and macroscopic plant remains were analyzed from a peat core representing the last two to three centuries of time collected in the ANSI part of the bog. The resulting pollen diagram (Figure 10) clearly contains pollen of trees such as *Pinus strobus, Tsuga, Betula, Ostrya/Carpinus, Quercus, Fagus, Carya,* and *Fraxinus* and of bog species such as Ericales and spores of *Sphagnum* in the pre-European part of the record.

The characteristic abundance of *Ambrosia* pollen, representing the European period, is evident in the top of the core. It is accompanied by the increased presence of pollen of other weed species such as Gramineae, *Rumex,* Chenopodiaceae, and *Artemisia*. The peat in this interval is highly humified and contains abundant charcoal layers.

Archival records indicate that the first European settler had claimed a parcel of land in 1802 about 15 km south of Wainfleet Bog on the shore of Lake Erie. However, settlement of any significance did not occur until the 1820s (Nagy, 1992). The first roads extended north of the Lake Erie shore settlement around 1829. Therefore, a date of around 1830 is assigned to the beginning of the European period, as indicated by the rise in *Ambrosia* pollen signifying forest clearance and expansion of agricultural fields in the vicinity of Wainfleet Bog.

A bog appears to have existed at the study site prior to 1830 but changed noticeably after the arrival of Europeans. The most obvious change appears to be a trend to a much drier surface. Hydrological processes were severely altered by the Europeans and this also caused changes to physical, geochemical, and biological processes in the system. Peat accumulation has been impaired as a result.

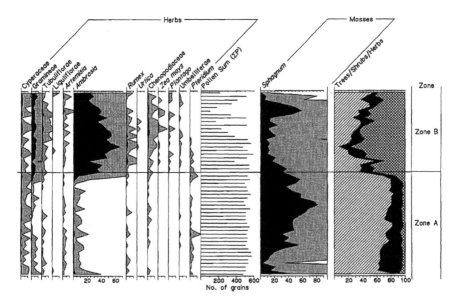

Figure 10 (continued)

Charcoal layers are common in the upper part of the peat, indicating that peatland fires were now more frequent than during the earlier period. The steady increase in *Betula* pollen at the surface is likely due to increases in growth of *Betula papyrifera* in and around the bog and to the arrival of *B. occidentalis* which is common on the bog today. The surface oxic layer has probably thickened compared to its pre-European thickness, allowing terrestrial species such as the *Betula occidentalis* to invade.

CHANGES IN VERTICAL GRADIENTS AND CLASSIFICATION OF PEATLAND CONDITION

The two examples presented above show changes in the vertical peat gradient reflected by pollen and spore analysis which can be related to forest clearance and other activities associated with historical European settlement as the primary cause. The impact appears to have altered long-term processes by intensifying rates that now may be much different than the pace set throughout most of their history of development prior to the arrival of the Europeans. This change in the natural processes in the peatland may go undetected; we may be falsely assuming that existing peatland systems are on their natural trajectory of development when in fact they clearly may not be.

Many of the existing peatlands in southern Ontario probably represent altered ecosystems. Owing to widespread forest clearing, modification for agriculture, draining, damming, and conversion of land for commercial, industrial, and residential purposes, peat lands currently are sustained by changed processes quite

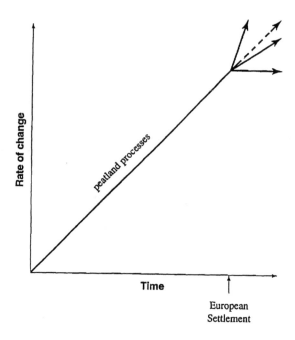

Figure 11 Schematic plot depicting the change to processes in southern Ontario peatlands as a consequence of European settlement. The processes giving rise to peat accumulation may decrease, such as in degraded peatlands, or may increase, such as in floating peatlands in kettle holes.

Table 1   Scheme for Classifying the Extent of Human Impact on Peatlands in Southern Ontario

NATURAL: These peatlands have not been influenced by human activity during the European period.

MINOR DISTURBANCE: Peatlands partly influenced by humans, but retaining nearly the same peatland type and form as before human disturbance occurred locally.

MODERATE DISTURBANCE: A functioning peatland system is retained but basin structure and form may have changed and the peatland has transformed or is in a state of transformation from one peatland form and type to another.

MAJOR DISTURBANCE: These peatland systems have been affected by human activity in such a way that basin structure and form have changed. Conditions at the site have led to a general trend of species impoverishment or replacement of the dominant species by another species.

ARTIFICIAL: These peatlands include those entirely conceived and created by humans simulating natural or seminatural peatlands in basin structure, function, and form.

different from those that existed before the arrival of Europeans (Figure 11). In an attempt to classify peatlands affected by human activity, a five-tier scheme has been devised (Table 1). This scheme attempts to generalize the degree to which human activities have affected natural processes within peatland systems. It is a modification of a more general system proposed by Westhoff (1983).

The Wainfleet Bog example shows the degree to which human activities can contribute to an overall degradation of a peatland system. Changes in peat decomposition processes and the expected decline in peat accumulation are probably some of the consequences of human changes to the peatland and its environment during the historic period. Further work is needed to determine more accurately which processes have been most impaired. This knowledge is vitally important to its future management. Restoration should be directed towards reversing undesirable trends by slowing down current processes.

The case of the floating peatlands is a good example of human activity contributing to net gains to peatlands (Gosselink and Maltby, 1990). These kinds of peatlands are particularly important from a management and conservation point of view. If an augmentation of water supplies was really responsible for initiating peat formation in kettle holes, would artificial flooding be a viable management option to rehabilitate degenerated peatlands in the region? If human activities can create peatlands by accident, can they be repeated intentionally through management? Much remains to be learned about the different ways peat forms along the vertical gradient and the mechanisms controlling it.

Following the scheme presented in Table 1, both of the examples discussed here represent seminatural or cultural peatland systems. In the case of the floating peatlands, European activities altered site hydrology which led to a net gain in peatland at the expense of open water wetland and lake. If this is viewed as positive, then measures for management need to be directed towards conserving the hydrological processes operating in these systems. Personal observation of some of these systems over the last 20 years is revealing changes in the biota apparently symptomatic of alterations in hydrological and geochemical processes over this short period of time. Is the average lifetime of a floating peatland 150 to 200 years? Conservation need not mean "hands-off". The case of the floating peatlands may represent a circumstance where the disturbance must be maintained to sustain the peatland. Is this what is needed for the floating peatlands since they were created by disturbance?

Wainfleet Bog is in critical need of restoration. A more concerted management plan is required to restore this bog. A scheme to reduce all human impacts, to conserve and augment water resources (hydrological processes), to create adequate vegetation buffers in the peatland and around the periphery (geochemical processes), and to facilitate regrowth and reestablishment of peatland plants (biological processes) is required.

ACKNOWLEDGMENTS

I thank L. R. Belyea, H. J. Kubiw, S. G. Marsters, and B. R. Nagy for contributing ideas and data presented here. Drs. R. Aravena, M. J. Bunting, E. McBean, and G. Mulamoottil kindly read early versions of the manuscript which helped to improve it substantially. This work is supported by grants from the Ontario Ministry of Natural Resources and the Natural Sciences and Engineering Research Council of Canada.

REFERENCES

Aario, L., 1932. Pflanzentopographische und palëographische mooruntersuchungen in N-Satakunta, *Fennia* 55:1–179.

Auer, V., 1927. Stratigraphical and morphological investigations of peat bogs of southeastern Canada, *Communicationes ex Instituto Quaestionum Forestalium Finlandiae Editae* 12:1–62.

Canadian National Wetlands Working Group, 1988. *Wetlands of Canada*, Polyscience Publications Inc., Montreal.

Clymo, R. S., 1983. Peat, in Gore A. J. P., ed., *Mires: Swamp, Bog, Fen and Moor, Ecosystems of the World 4A*, Elsevier, Amsterdam, pp. 159–224.

Clymo, R. S., 1991. Peat growth, in Shane, L. C. K., and Cushing E. J., eds., *Quaternary Landscapes*, University of Minnesota Press, Minneapolis, pp. 76–112.

Clymo, R. S., 1992. Productivity and decomposition of peatland ecosystems, in Bragg O. M., Hulme, P. D., Ingram, H. A. P., and Robertson, R. A., eds., *Peatland Ecosystems and Man: An Impact Assessment*, University of Dundee, Dundee, pp. 3–16.

Cowardin, L. M., Carter, V., Golet, F. C., and Laroe, T., 1979. Classification of wetlands and deepwater habitats of the United States, *U.S. Fish and Wildlife Service* FWS/OBS-79/31:1–131.

Damman, A. W. H. and French, T. W., 1987. The ecology of peat bogs of the glaciated northeastern United States: A community profile, *U.S. Fish and Wildlife Service Biological Report* 85(7.16): 1–100.

Fraser, J., 1989. The Wainfleet Bog: A recent example of government and private agency cooperation in wetland conservation and management, in Bardecki, M. J. and Patterson, N., eds., *Wetlands: Inertia or Momentum*, Federation of Ontario Naturalists, Don Mills, Ontario, pp. 135–141.

Gentilcore, R. L. and Head, C. G., 1986. *Ontario's History in Maps*, University of Toronto Press, Toronto.

Gorham, E. and Janssens, J. A., 1992. Concepts of fen and bog re-examined in relation to bryophyte cover and the acidity of surface waters, *Acta Societatis Botanicorum Poloniae* 61:7–20.

Gosselink, J. G. and Maltby, E., 1990. Wetland losses and gains, in Williams, M., ed., *Wetlands: A Threatened Landscape*, Blackwell, Oxford, pp. 296–322.

Göttlich, K., 1990. *Moor- und torfkunde*, E. Schweizerbart'sche Verlagsbuchhandlung, Stuttgart.

Ingram, H. A. P., 1978. Soil layers in mires: Function and terminology, *Journal of Soil Science* 29:224–227.

Ivanov, K. E., 1981. *Water Movement in Mirelands*, Academic Press, London.

Johnston, C. A. and Naiman, R. J., 1987. Boundary dynamics at the aquatic–terrestrial interface: the influence of beaver and geomorphology, *Landscape Ecology* 1:47–57.

Kulczynski, S., 1949. Peat bogs of Polesie, *Memoires Academiae Polesie Sciences* B15:1–356.

Leverin, H. A., 1941. Peat moss deposits in eastern Canada, *Department of Mines and Resources, Memorandum Series No. 80*, 1–49.

Moore, P. D., 1987. Ecological and hydrological aspects of peat formation, in *Coal and Coal-Bearing Strata: Recent Advances*, Scott, A. C., ed., Geological Society, Special Publication 32, London, pp. 7–15.

Nagy, B. R., 1992. Postglacial Paleoecology and Historical Disturbance of Wainfleet Bog, Niagara Peninsula, Ontario. M.A. thesis, University of Waterloo, Ontario.

Overbeck, F., 1975. *Botanisch-geologische moorkunde*, K. Wachholz Verlag, Neumünster.

Rowell, T. A., 1988. *The Peatland Management Handbook*, Nature Conservancy Council, Peterborough, U.K. Publication 14.

van Dierendonck, M. C., 1992. Simulation of peat accumulation: An aid in carbon cycling research? *Suo* 43:203–206.

von Bülow, K., 1929. *Allgemeine moorgeologie*, Verlag von Gebrüder Bortraeger, Berlin.

Warner, B. G., 1993. Palaeoecology of floating bogs and landscape change in the Great Lakes drainage basin of North America, in Chambers, F. M., ed., *Human Impact on the Landscape and Climate Change*, Chapman and Hall, New York, pp. 237–245.

Warner, B. G. and Bunting, M. J., 1995. Indicators of rapid environmental change in northern peatlands, in Berger, A. R., ed., *Geoindicators of Rapid Environmental Change*, A. A. Balkema, Rotterdam.

Warner, B. G., Kubiw, H. J., and Hanf, K. I., 1989. An anthropogenic cause for quaking mire formation in southwestern Ontario, *Nature* 340:380–384.

Westhoff, V., 1983. Man's attitude towards vegetation, in Holzner, W., Werger M. J. A., and Ikusima, I., eds., *Man's Impact on Vegetation*, Dr. Junk, The Hague, pp. 7–24.

Williams, M., 1990. Protection and retrospection, in Williams, M., ed., *Wetlands: A Threatened Landscape*, Blackwell, Oxford, pp. 321–353.

URBAN INTENSIFICATION AND ENVIRONMENTAL SUSTAINABILITY: THE MAINTENANCE OF INFILTRATION GRADIENTS

5

E. A. McBean, G. Mulamoottil, Z. Novak, and G. Bowen

ABSTRACT

The trends toward urban intensification and the implications to adjacent wetlands in response to the changes in the quantity and quality of runoff are considered. Four urbanization scenarios are examined as they influence inputs to wetlands; these are characterized by changes in quantity, frequency of inputs and the pathways of migrating water. The findings indicate that the mitigation of changes to the hydrologic balance, as it relates to the maintenance of gradients for ground water recharge, will require significant costs in dollars and space.

INTRODUCTION

According to Lowe (1991), historically the rate of growth in cities in the industrial countries has been diminishing, and even with declining populations in the core sections the urban areas are continuing to move outward. For example, in North America and Australia zoning codes are being enforced to restrict residential densities and to minimize urban expansion. However, this move has achieved the opposite effect because, with reduced densities of residential zoning, developers are forced to utilize more land. As new communities emerge, transportation services expand, which encourages further urban sprawl.

The way in which cities utilize land and resources in turn influences the environment. To protect the environment there is a need to utilize land resources in response to the urban sprawl syndrome and to recognize the carrying capacity of the biophysical environment. If the environment is not protected, resources upon which current and future societies depend will be destroyed (Lowe, 1991).

In Canada, and especially in southern Ontario, urbanization is consuming valuable agricultural land and is encroaching upon environmentally sensitive areas and wetlands at a rapid pace. Without appropriate mitigation, these natural landscapes will be seriously transformed and changed.

One element of the overall environmental impact of development activities in urban areas is the change in ground water infiltration levels. The focus of the research described in this chapter is to examine the changes in the infiltration gradients, since any changes to the infiltration gradients will modify the hydrologic regime which will then influence the long-term viability of wetlands of an area. If the predevelopment infiltration gradients are not maintained in the postdevelopment stages, the functions and, therefore, the sustainability of wetlands are certainly in question.

Assuming that urban development standards are changing and evolving, changes in environmental gradients should be minimized. This will reduce impacts on wetlands and provide some insights as to how changing gradients may influence migration pathways of pollutants that will impinge upon wetlands. The objective of this chapter is to examine the degree to which infiltration gradients for ground water recharge can be maintained under development scenarios currently being utilized in southern Ontario.

CHANGES IN URBAN DEVELOPMENT STANDARDS

To meet the future housing needs in urban centers, while allowing for the effective use of land, numerous planning initiatives, policy statements, and consultant documents were generated during the last 15 years. By 1990, there seems to have emerged a consensus for the potential for housing intensification to provide social (affordable housing), economic (lower infrastructure costs), and environmental benefits. Housing intensification is defined here as "residential development of a site at a density that is substantially higher than previously existed or designated for that kind of site." The term housing intensification refers to conversion, infill, redevelopment, adaptive reuse, and suburban densification (Canadian Urban Institute, 1991).

Policies, constraints, and options related to housing intensification are outlined in a report by the Canadian Urban Institute (1991). The Royal Commission on the Future of the Toronto Waterfront (Crombie, 1991) has pointed out the inadequacies of the form and pattern of urban growth which are influenced by housing standards. The standards such as lot sizes, setbacks, and road widths affect the amount of land consumed. The Commission has been particularly critical of the housing standards that are well entrenched in the municipal planning and development approval process. Due to the reluctance of municipalities to respond to developers' requests to reduce the size of lots, it is difficult to provide affordable housing and compact communities. In "Making Choices: Alternative Development Standards — Guideline" by Marshall *et al.* (1994), an attempt has been made to combine development standards to create new kinds of streets and neighborhoods. These standards are expected to provide cheaper housing and efficient use of the land.

An important aspect of the trend toward intensification is the relentless pressure from the developers to accommodate more housing units. With urban intensification, there will be energy savings and improvements of the ambient air quality of an area, greater use of public transit facilities, and a decrease in the use of private automobiles. At the same time, among policy makers, there are strong interests in improving the quality of the environment and reducing the negative impacts of urbanization. The justification for imposing higher environmental standards is based on the need to achieve environmental sustainability in urban areas, and a concern to minimize the impacts of an individual subdivision on surrounding areas, particularly municipalities downstream.

BEST MANAGEMENT PRACTICES FOR WATER QUANTITY AND QUALITY CONTROLS

In the early 1970s stormwater management (SWM) controls primarily focused on water quantity to provide protection of life and property. Over time, new dimensions of SWM controls were instituted to minimize water quality degradation in downstream areas. Historically, most of the best management practice (BMP) options were "end-of-the-pipe" controls such as the construction of wet or dry ponds to contain the stormwater runoff from a subdivision. However, with the acceptance of the concept of environmental sustainability, changes to the "end-of-the-pipe" philosophy are imminent. The use of a combination of dispersed control measures scattered throughout the subdivision are necessary, and they are schematically depicted in Figure 1. The purpose of the dispersed control measures is to make the components function as a system, each assisting toward accomplishment of the overall objective. The type of BMPs indicated in Figure 1 are now having a significant influence in modifying subdivision design, as for example in the River Oaks subdivision in Oakville, Ontario.

In considering an array of BMP options, some of the alternatives will decrease the source generation of excess runoff and pollutants. The reduced use of directly connected pavement does precisely that. Some of the other options modify the hydrograph either during the runoff processes themselves (e.g., infiltration trenches encourage infiltration during transit) or just prior to discharge to the receiving water body (e.g., a wet pond provides temporary storage and thus attenuates the peak runoff condition). Thus, if the BMPs are allowed in a subdivision which is established around a wetland, there will be changes in the horizontal gradients of infiltration, from the predevelopment to the postdevelopment conditions. Further, alternative BMP options will modify the infiltration regime in different ways.

The changes in land use influence the source generation levels of pollutants. The application of BMPs, besides creating changes to the infiltration gradients, will also potentially change the amounts of pollutants entering wetlands. Another

Individual Lots

- encourage minimal use of directly–connected impervious areas

- provide backyard swales

- reduce lot size

- educate residents on the use of chemicals, oils and fertilizers and in disposal of residues

- efficient de–icing programs

- proper disposal of pet droppings

Conveyance and Pre–treatment Possibilities

- use of swales, perforated pipes, and infiltration trenches

- use of modular block pavement

- elimination of curbs and gutters

Pre–treatment Opportunities

- use of infiltration basins

- use of vegetated buffer areas

- use of off–line ponds

Final Treatment and Attenuation Options

- use of retention and detention ponds

- use of constructed wetlands

Figure 1   Examples of dispersed controls being considered for implementation in a subdivision.

important factor that determines the quantity of pollutants that will reach the wetlands is the mode of conveyance of the pollutants. When the changes in the infiltration gradients occur in such a way that larger surface-water runoff occurs due to decreased infiltration and ground water movement, attenuation processes such as filtration, adsorption, and dispersion, which act to decrease the pollutant concentrations, become limited.

CHARACTERISTICS OF WETLANDS

The importance of the hydrologic regime to wetland ecology cannot be overemphasized. Indeed, hydrology is the driving force of a wetland and intermit-

Table 1   Changes in Water Regime of Wetlands

I. Impacts associated with surface flows
   Changes in mean water level
   Changes in periodicity of water level fluctuations
   Changes in flow circulation patterns
II. Impacts associated with ground water flows
   Changes in local water table levels
III. Impacts associated with *creation* of channels in wetlands
   Drainage of surface waters
   Elimination of periodic flooding
   Changes in retention storage
IV. Impacts associated with water quality
   Fertilization as a result of urban runoff inputs
   Contributions from urban runoff including dimensions associated with turbidity,
      chemical pollution, and temperatures

tent changes in the hydroperiods of wetlands influence species composition and productivity. Many wetland plants are adapted to a limited range of water depths and some plants can tolerate continuous inundation and/or drying for only limited periods of time. As a result, if urbanization influences the hydroperiod, which in the first place is responsible for the development and maintenance of a wetland, the ecology of the wetland will change and potential diversity and the viability of the wetland in its natural state may be lost.

The prediction of the response of a constructed or natural wetland to the storm water influent regime of an urbanized area is a significant challenge. In urban areas where land is expensive, even though the opportunities for use of constructed wetlands for flow attenuation are very limited, the major reason for the inclusion of wetlands as a BMP is attributed to their ability for contaminant attenuation. Many questions remain unanswered regarding the long-term viability of the use of natural or constructed wetlands as a BMP and even of the appropriateness for their use. A wetland is more efficient than a wet pond for pollutant removal, but the amount of area required to establish a constructed wetland is considerably greater (P'ng et al., 1993).

The problems of maintaining wetlands on a sustainable basis are much greater for small wetlands. Given the rapid progression and commitment toward urban densification, the sustainability of these small wetlands is even more in doubt. Examples of the changes in the water regime of wetlands related to storm water inputs are listed in Table 1.

IMPACTS ON GRADIENTS

As an essential element toward estimating the effect of urbanization on adjacent wetlands, the present study was undertaken to determine the degree to which the infiltration gradients associated with wetlands could be maintained at preurbanization levels. As part of this research, a detailed water balance model (HELP, after Schroeder et al., 1988) was used to quantify daily levels of the various components of the water regime, including those of evapotranspiration, surface runoff, changes in soil moisture, and ground water infiltration responses

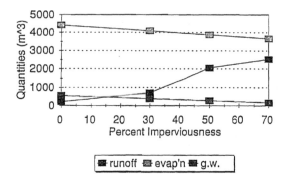

Figure 2   Alternative percentages of runoff, evaporation, and ground water recharge in response to imperviousness (seasonal time-frame, June 1 to September 30).

under different surface cover conditions. The model includes as examples of the input data the solar radiation, temperature, precipitation, surface abstraction depth, and depth to the vegetative root zone in the case of permeable surfaces. The model also integrates over a specified time-history of meteorologic conditions (see Figure 2). Thus, by changing the features of the watershed, as for example, the percentage of contributing land that is impervious, it is possible to quantify the changes in infiltration quantities as a result of changes in imperviousness of the contributing drainage area.

CHANGES IN GROUND WATER RECHARGE AND SURFACE WATER RUNOFF QUANTITIES AS A RESULT OF URBANIZATION

Four scenarios of urbanization involving different degrees of imperviousness which might influence ground water recharge in southern Ontario are examined. The results are analyzed and summarized in Table 2(a). These results show the changes to ground water recharge levels in response to changes in the imperviousness in the contributing drainage area. As shown in Table 2(a), the ground water recharge decreases to 35% of predevelopment conditions when the percentage of contributing impervious land is increased to 70%. In Table 2(b), the changes in surface runoff volumes in response to changes in impervious surfaces are summarized. As indicated by the results in Table 2(b), the surface runoff increases by 14.7 times as a result of a change in imperviousness to 70% of the contributing drainage area. The implications of these findings on wetland hydrology are as follows:

1. As a result of urbanization, total water quantity inputs to an adjacent wetland are much larger. The predevelopment inputs consist of ground water recharge of 28,097 m³ and additional surface water inputs of 19,106 m³, for a total input volume of 47,203 m³ to the wetland. This predevelopment input volume to the wetland is considerably smaller than that associated with a 70% development in which ground water input is 9764 m³ in addition to surface water inputs of

Table 2(a)   Influence of Varying Degrees of Imperviousness on Ground Water Recharge

		Ground water recharge	
Scenario	Percentage imperviousness	Quantity (m³)	Relative to predevelopment conditions
I	Predevelopment	28,097	1
II	30%	20,238	0.72
III	50%	15,001	0.53
IV	70%	9,764	0.35

Table 2(b)   Influence of Varying Degrees of Imperviousness on Surface Runoff

		Surface water runoff	
Scenario	Percentage imperviousness	Quantity (m³)	Relative to predevelopment conditions
I	Predevelopment	19,106	1
II	30%	131,100	6.9
III	50%	205,758	10.8
IV	70%	280,416	14.7

280,416 m³, for a total of 290,280 m³. The result of the urbanization represents an increased water input of 610% to the wetland. The increase in runoff to the wetland is directly attributable to surface water since the ground water recharge has decreased by 35% from predevelopment levels. Further, with higher levels of "impervious" areas there is a much lower volume of water that goes into the atmosphere through evapotranspiration.

2. The pathways by which the water reaches the wetlands are also altered considerably as a result of the increase in the percentage imperviousness associated with urbanization. As noted above, a decrease in ground water recharge occurs, but in terms of total quantity this decrease is more than compensated for by a considerable increase in surface water runoff. As most of the water is moving through surface water runoff, the contaminant attenuation that could be expected during ground water infiltration by processes such as adsorption, filtration, and biodegradation will be lost. The modification of the hydrological regime impacts on the wetlands in terms of both the volume of inflowing water and on the entrainment of water quality constituents. The wetlands will thus receive substantial increases in such constituents as suspended solids, organics, greases/oil, phosphorus, and coliforms. The maximum impact on the wetland occurs when the imperviousness is 70%. There is a similar trend in the contaminant transport with lesser impervious areas.

3. The frequency of "pulsed" inputs arising from the surface runoff is also greatly increased with urbanization. As an indication of the magnitude of the changes, the number of days in which surface water runoff occurs for different impervious areas is indicated in Table 3. The values in Table 3 are for five separate years, to demonstrate the impact of different years where the timing and size of storms varies from year to year. Many more days of nonzero surface runoff to the wetland occurs due to the lower water-holding capacity of water abstraction

Table 3   Number of Days of Nonzero
Surface Runoff to the Wetland
for Different Impervious Cover

	Number of days of nonzero surface runoff[a]	
Year	Predevelopment condition	Impervious cover
1	10	86
2	16	87
3	14	114
4	12	97
5	7	73
Average	11.8	91.4

[a] Using an annual time-frame.

processes for the impervious areas, in comparison with the relatively high water-holding capacity of pervious areas. The ratio shown in Table 3 indicates that there are, on average, 7.7 times as many days with nonzero runoff from the 70% impervious land use than days with nonzero runoff from the predevelopment conditions.

The increased frequency and higher volumes of surface water runoff with elevated pollution levels are a direct consequence of urbanization. The long-term sustainability of a wetland in urban areas is thus open to question.

POTENTIAL FOR INFILTRATION

With the Province of Ontario encouraging urban intensification which will result in more impervious areas, it is important to explore whether the potential exists to maintain or approach more closely the predevelopment infiltration gradients. If pre- and postdevelopment infiltration gradients are similar, then the pollutant attenuation processes will be more effective and the inundation of the wetland will be less frequent. Based on the results provided in Table 2(a), the loss of infiltration as a result of varying degrees of imperviousness are noted in the third column of Table 4. Taking this calculation one step further, the effectiveness of various sizes of detention ponds for a 1-ha drainage area was examined, and the results are presented in Figure 3. As an example, reading from Figure 3, a detention basin of 1000 m^3/(ha of contributing drainage area) would not capture 740 m^3/(ha of drainage area) of surface runoff when there is 30% imperviousness. If the imperviousness percentage is increased to 70%, then the runoff volumes not captured correspond to 3150 m^3, which is a considerable increase from the predevelopment condition.

The results from Figure 3 can, in turn, be utilized to approximate the size of detention volume needed to match the predevelopment levels of ground water recharge. The values in Table 4 correspond to those for the River Oaks subdivision in Oakville, Ontario which has a drainage area of 6.9 ha. As listed in Table 4,

Table 4   Necessary Detention Volumes to Match Predevelopment Ground Water
                Recharge Volumes[a]

Scenario	Percentage imperviousness	Increased groundwater recharge needed[b] (m³)	Needed detention volume to match predevelopment groundwater recharge conditions (m³)	Surface area of infiltration basin[c] (%)
I	Predevelopment	0	0	0
II	30%	7,859	4,485	4.3
III	50%	13,096	6,555	6.3
IV	70%	18,333	10,005	9.7

[a] Assuming the site as 6.9 ha in contributing drainage area.
[b] Needed to match predevelopment conditions.
[c] Assuming the maximum depth of water to be 1.5 m, and the surface area as a percent of the contributing drainage area.

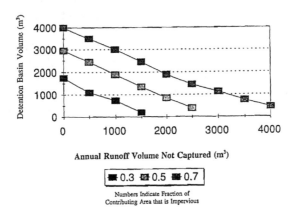

Numbers Indicate Fraction of
Contributing Area that is Impervious

Figure 3   Trade-offs between detention volume and runoff volume not captured, for varying degrees of imperviousness.

anadditional ground water recharge of 7859 m³ is needed for the 6.9-ha site, for 30% imperviousness to match predevelopment conditions. Thus 1139 m³/ha needs to be infiltrated to meet predevelopment conditions. To capture 1139 m³/ha requires (as read from the 30% line on Figure 3) approximately 650 m³/ha of detention volume is needed to enable the infiltration to occur. Therefore, for a drainage area of 6.9 ha, a detention volume of 4485 m³ (650 m³/ha × 6.9 ha drainage area) are required. This magnitude of detention area required has serious implications. For example, assuming an impoundment with a depth of 1.5 m, 4.3% of the drainage area in the form of infiltration basins will be required. As a continuation of this line of thinking (fifth column of Table 4) for varying degrees of urbanization, sizable infiltration areas are required — up to 9.7% of the drainage area when the imperviousness reaches 70%.

The inference of these findings is that the infiltration to the ground water regime needed to maintain the predevelopment infiltration gradient for the postdevelopment conditions represents a substantial challenge. Maintaining the gradient of ground water infiltration to a wetland will require significant infiltration needs and, as a direct result, ongoing maintenance activities to ensure that infiltration rates continue to be maintained. It is important to point out that the likelihood of attaining environmental sustainability would be greatly improved if the infiltration is maintained at numerous points throughout the contributing drainage shed and not concentrated in close proximity immediately adjacent to a wetland.

CONCLUSIONS

The trend toward urban intensification will continue because of the increasing population and concerns due to monetary costs related to infrastructure and the environmental costs associated with urban sprawl. To mitigate the detrimental changes to the hydrologic balance, particularly as it relates to the maintenance of gradients for ground water recharge toward wetlands, significant costs in dollars and space will have to be incurred. The mitigation measures should be spread throughout the contributing drainage areas. From an environmental sustainability perspective, small-scale infiltration facilities seem to be more acceptable as large facilities will have major localized impacts.

REFERENCES

Canadian Urban Institute, 1991. *Housing Intensification: Policies, Constraints and Options*, Urban Focus Series, 90-6, Toronto, Canada.

Crombie, D., 1991. *Regeneration: Toronto's Waterfront and the Sustainable City: Final Report*, Royal Commission on the Future of the Toronto Waterfront, Toronto.

Lowe, M. D., 1991. Shaping cities: The environmental and human dimensions, *Worldwatch Paper*, 105.

Marshall, Macklin, and Monaghan, Ltd., 1994. *Making Choices: Alternative Development Standards — Guideline*, Ontario Ministry of Housing and Ontario Ministry of Municipal Affairs, Toronto. 177 pp.

P'ng, J., Bryant, G., and Andrews, D., 1993. Ontario Stormwater Quality Best Management Practices (BMP) and Design Manual, presented at Sixth International Conference on Urban Storm Drainage, J. Marsalek and H. Torno, eds., Niagara Falls, Ontario, September 1993.

Schroeder, P. R., Gibson, A. C., and Smolen, M. D., 1988. *Hydrologic Evaluation of Landfill Performance (HELP) Model.* Vol. 2, Documentation for Version 2, U.S. Environmental Protection Agency.

ROOT ZONE MOISTURE GRADIENTS ADJACENT TO A CEDAR SWAMP IN SOUTHERN ONTARIO

6

D. E. Stephenson and D. B. Hodgson

ABSTRACT

Subsurface moisture levels were monitored at three depths within the rooting zone adjacent to, and within, a cedar (*Thuga occidentalis*) swamp in Kitchener, Ontario. Soil moisture readings, rainfall, and seepage outflow to a pond were measured on 85 days over a 4-month period to evaluate fluctuations in soil moisture around the swamp. The impact of shallow subsurface water flows in the root zone and the nature of land use changes in the catchment basin were investigated. The implications of moisture level fluctuations in upland stands of trees adjacent to wetlands were examined, with special reference to provincial wetlands delineation criteria and setbacks.

INTRODUCTION

Numerous jurisdictions in North America have developed guidelines for delineating wetland boundaries using characteristics of the vegetation, soils, and hydrology. The U.S. has prepared lists of wetland indicator plant species for delineation (Tiner, 1991). The evaluation system produced by the Ontario Ministry of Natural Resources (OMNR) relies heavily on vegetation species composition to determine the boundary of a wetland, using the point where 50% of the plant community consists of a mix of wetland and upland species (Ontario Ministry of Natural Resources, 1993).

In the Ontario Ministry of Natural Resources' (1993) evaluation system, wetlands are defined as: "Lands that are seasonally or permanently flooded by shallow water as well as lands where the water table is close to the surface; in

1-56670-147-3/96/$0.00+$.50
© 1996 by CRC Press, Inc.

either case the presence of abundant water has caused the formation of hydric soils and has favored the dominance of either hydrophytic or water tolerant plants."

Although hydric soil is not defined explicitly in the OMNR system, hydric soil is soil that is generally saturated, flooded, or ponded long enough during the growing season to develop anaerobic conditions in the upper portion and is developed under conditions sufficiently wet to support the growth and regeneration of hydrophytic vegetation (Tiner, 1991). Hydrophytic vegetation includes plants able to grow in water or in substrates that are periodically deficient in oxygen (anaerobic) during a growing season as a result of excessive water content (Tiner, 1991).

Root soil moisture is important because many plant species found in wetlands have broad tolerances and may inhabit a range of soil conditions. Eastern White Cedar (*Thuja occidentalis*) has proven to be problematic since this species is found in both upland and wetland sites (Ontario Ministry of Natural Resources, 1993). On the other hand, Golet and Lowry (1987) found that soil water regime accounted for approximately 50% of the variation in *T. occidentalis* growth.

Golet and Lowry (1987) stated that monitoring ground water levels is an appropriate approach for estimating wetland water regimes, but is only a crude measure of soil wetness. Soil moisture data from various depths within the rooting zone are required to develop a better understanding of the relationships between tree growth and soil wetness.

Since the amount of water in the soil directly affects the growth of plants, the determination of water content is an easily measured soil parameter. However, it is not an independent variable and many measures of soil moisture are difficult and do not necessarily have direct ecological meaning (Carter, 1986; Topp, 1993). Soil moisture can be estimated by calculating moisture content per unit mass or by using time domain reflectometry. The estimation based on mass is the common procedure for determining the water content of soils and typically consists of comparing the *in situ* weight and oven-dried weight. The loss in weight is expressed as a percentage of oven-dried weight. This testing procedure is time consuming, especially if multiple repetitions are conducted over time. Oven drying of soil samples with high organic matter contents can also result in mass losses arising from volatilization of organic components (Westerman, 1990; Topp, 1993).

Time domain reflectometry (TDR) is a method of determining volumetric soil water content. This technique makes use of the electrical properties of water molecules to determine the water content of soil (Topp, 1993). TDR measures the velocity of the propagation of a high-frequency signal through the soil. This method provides a measurement of the average water content of the soil situated between parallel probes. By determining the travel time of an electric pulse through the soil between the cables or probes, the dielectric constant of the soil can be calculated and then applied to an equation to give the volumetric water content. Water contents at various depths can be determined by using a series of probes positioned horizontally in the soil at the required depths (Pringle, 1995).

The soil probes are connected to wires which extend to the soil surface. The bundled wires may be left in place indefinitely.

The water content of a soil is a constantly changing parameter which is dependent on numerous factors such as soil texture, structure, position in the landscape, drainage, and vegetation cover. To characterize soil water content over a period of time and to relate this information to wetland ecology is a challenge. Research has focused on the effects on plants of increased moisture content, especially flooding, and reductions in soil moisture. Several factors affect the availability of the soil water for plants, for example, soil texture. Many methods for determining soil water do not take these factors into account, and therefore may provide precise measures of soil water but are of less value in ecological research.

The soil tensiometer is described as an accurate technique for determining the relative wetness of soil (Harris, 1983). A soil tensiometer measures soil suction, which reflects the difficulty plant roots have in extracting moisture from the soil. Generally, a low soil suction value indicates high soil moisture. A soil suction reading of 0 cb (centibars) indicates saturated soil.

Soil suction is not easily converted to percent soil moisture values due to the impact of soil texture on moisture and the nonlinear relationship between soil suction and moisture. A 10 cb increase in soil suction around the saturation level would have a great impact on plants, whereas a 10 cb increase above 25 cb would have less impact.

This study focused on monitoring soil moisture fluctuations in the rooting zone (15, 30, and 45 cm below the surface) of white cedar (*Thuja occidentalis*) within a cedar-dominated swamp and an adjacent upland cedar stand. A soil tensiometer was used in order to investigate the feasibility and value of soil suction determinations.

METHODS

The study site was located in southern part of the City of Kitchener, Ontario adjacent to the Grand River west of Highway 8 (see Figure 1).

Soil moisture was monitored at 13 stations within the swamp and adjacent upland stand (see Figures 1 and 2). Upland stations were selected to maximize monitoring points at elevations between 0 and 1.5 m above the wetland boundary (Stations C and D). Two stations were selected higher on the slope (A and B: 2 and 2.5 m above the wetland boundary elevation, respectively).

Station E was selected on the wetland boundary, while F and G were located approximately 1 m lower in elevation than the wetland boundary. Station F was also established approximately midway between the wetland boundary and the pond.

Monitoring was carried out over 85 days between June 16 and October 10, 1990. Soil moisture was measured directly in the field at depths of 15, 30, and 45 cm at each station using a Soilmoisture Tensiometer Probe (model 2900F1 Soil

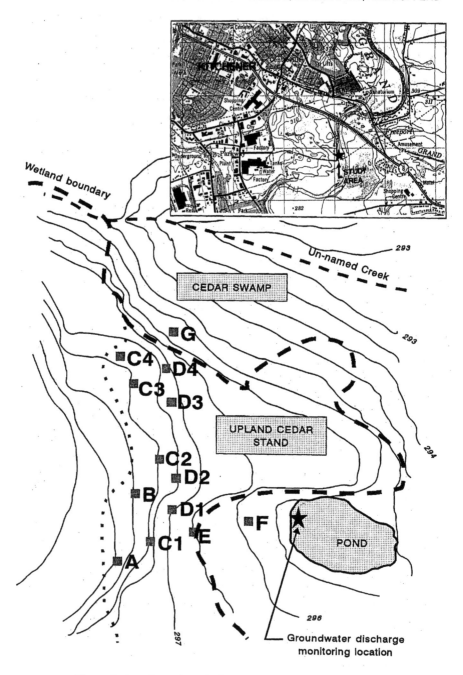

Figure 1   Locations of monitoring points (0.5 m contour intervals).

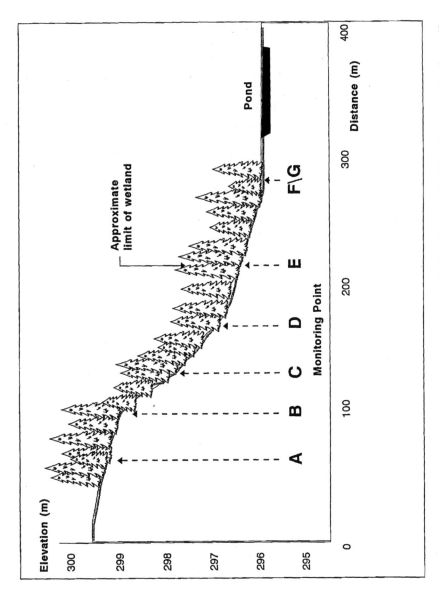

Figure 2   Schematic cross section through monitoring points (wetland is approximately at the same elevation as F).

Equipment Corporation, 1987). Since rooting depth of the cedars in both the upland and lowland areas was found to be no deeper than 30 cm, the monitoring depths represented: mid-rooting (15 cm), limit of the rooting zone (30 cm), and below the rooting zone (45 cm).

A pipe had been installed to monitor ground water flowing through the wetland into a man-made pond. An index of the subsurface ground water flow was obtained by monitoring discharge to this pond at the edge of the wetland.

The discharge volume was recorded during each of the 85 soil moisture monitoring days. This discharge volume was used to indicate fluctuations in the subsurface flows through the wetland. Precipitation measurements were obtained from a local weather station at the Waterloo-Wellington Airport.

Inspections of the soil in the area indicated that medium brown sand predominates down to depths of at least 100 cm.

The vegetation within the study area is predominantly white cedar stands which straddle the wetland boundary. The delineation of the wetland was based on understorey herbaceous plant species, as well as topography. The wetland boundary corresponded to approximately 297 m (amsl). The study area included a gradient between upland and lowland conditions. No changes in vegetation composition were observed over the elevation gradient studied (approximately 295.5 to 299.0 m amsl).

RESULTS

The results of monitoring soil suction in the study area are shown in Figures 3a and 3b and summarized in Table 1.

Ground water discharge flows were not significantly correlated (a <0.05) with rainfall, but the presence of peaks in ground water flows appeared in some cases to lag behind rainfall events (Figure 3b).

DISCUSSION

Although soil moisture fluctuations were not significantly correlated with rainfall or subsurface flows, soil moisture levels appeared to be related to fluctuations in both rainfall on the site as well as rainfall within the upslope catchment basin. Subsurface flows reflected rainfall, but also exhibited a short lag period during which upslope drainage occurred.

Soil was saturated during the June to October period at the 45 cm depth, which was deeper than the roots of the cedars. Saturation within the roots of the cedars was not observed at all wetland monitoring stations.

Stations A and B exhibited soil moisture content that differed from the other monitoring stations. Soil suction levels were typically high throughout the study period. The variability of soil moisture at the 45 cm depth, especially at Station A, was unlikely to affect the cedars since roots did not extend to this depth. The higher soil suction at the 45-cm depth at Station A may be due to the presence of coarse well-drained soils. This level may also have been above the majority of subsurface flows.

Stations A and B appear to be upland sites, however, Station G, with its constantly saturated soils, is a wetland site. The differences between Stations C, D, E, and F are more subtle. At the 45 cm depth, there is little or no difference between these stations. At the 15 and 30 cm depths, soil suction decreased from C to D to E, however, there was little difference between the soil suction readings at C and F.

Stations C through F extended over a horizontal distance of 150 m and represented a 2.0 m range in elevation. At these stations, vegetation, soil charac-teristics, and soil suction were generally the same.

The similarity between Stations C through F raises two important points:

1. The delineation of the wetland boundary does not appear to be precise or defensible in terms of vegetation, soil types, or soil moisture, and
2. In terms of sensitivity to soil moisture changes, Station C appears to be no different than F.

The two points are critical when considering the importance of boundaries and the protection of the functions of provincially significant wetlands according to the wetlands policy statement.

The subsurface flow of ground water was noted to be an important determi-nant of soil moisture at the middle and lower elevations. This suggests that soil moisture conditions, typical of wetland habitats and capable of supporting hydro-phytic vegetation, may extend considerable distances upslope from lowland wet-lands even in well-drained soils. The question of amount, duration, and frequency and seasonality of high soil moisture must be considered.

Soil tensiometers are relatively inexpensive and have proven useful for monitoring root zone moisture fluctuations in wetlands and adjacent habitats. This technique is a valuable tool that could yield critical ecological information which qualifies data used in conjunction with more traditional ground and surface water monitoring to assist with understanding soil moisture fluctuations both within and adjacent to wetlands.

STATION A

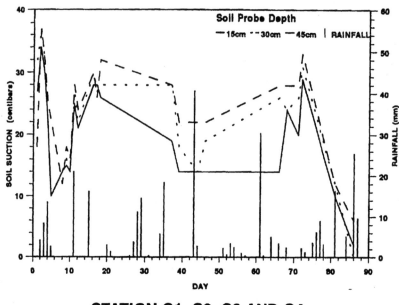

STATION C1, C2, C3 AND C4

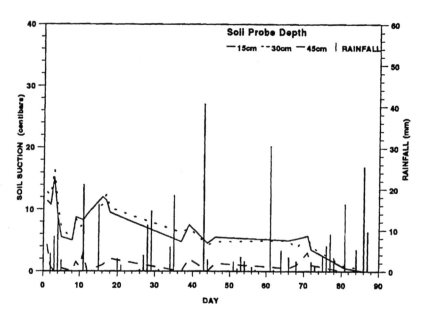

Figure 3a Graphs of soil suction at sample points A, B, C, and D on the 85 monitoring days.

STATION B

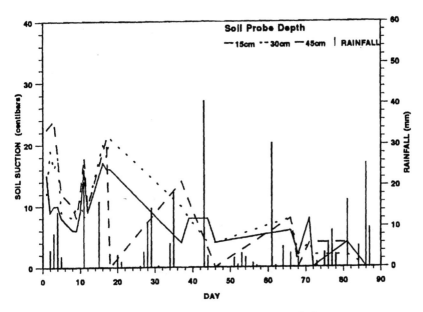

STATION D1, D2, D3 AND D4

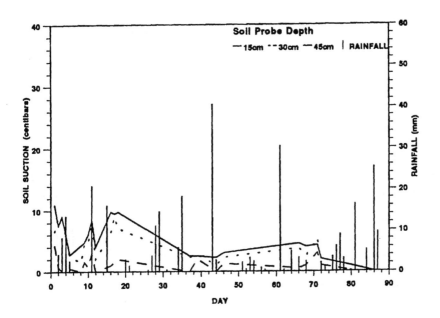

Figure 3a   (continued)

STATION E

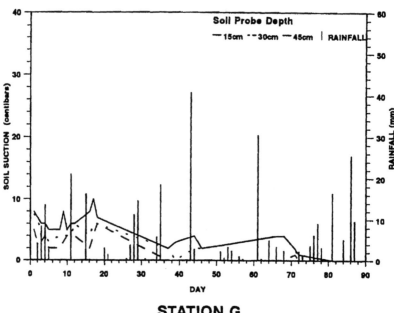

STATION G

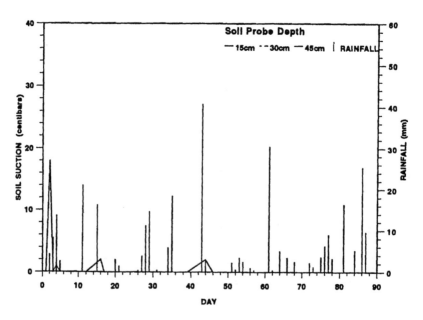

Figure 3b Graphs of soil suction at sample points E, F, and G as well as ground water flows on the 85 monitoring days.

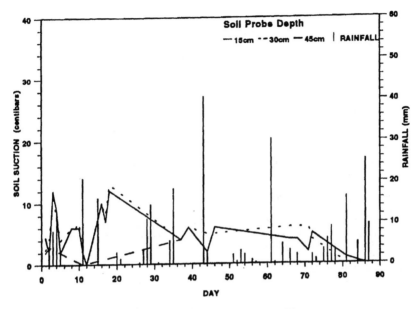

GROUNDWATER FLOWS

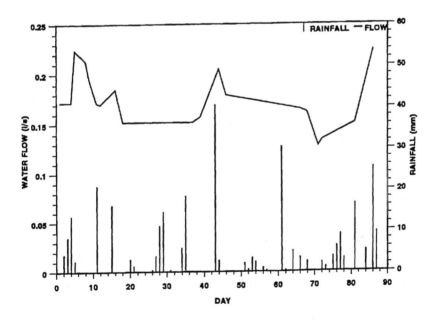

Figure 3b (continued)

Table 1   Summary of Soil Suction Monitoring

Station A
  Approximately 2.5 m above wetland boundary elevation
  Soil suction was generally greater than 12 cb at all depths
  As overall soil moisture levels increased (as shown by decreased soil suction values),
    especially in later part of the study period, soil moisture at the three monitoring depths
    was essentially the same
  During dry periods, deepest soils were the driest
Station B
  Approximately 2.0 m above wetland boundary elevation
  Considerable fluctuation in soil moisture was observed at this station
  At 15 and 30 cm depth soil suction decreased through study period
  At the 45 cm depth, soil saturation occurred periodically
Station C
  Approximately 1.0 to 1.5 m above wetland boundary elevation
  At the 45 cm depth the soil was close to saturation throughout the study
  Moisture levels at the 15 and 30 cm depths were generally between 5 and 10 cb and
    decreased throughout the study period
Stations D
  Approximately 0.5 to 1.0 m in elevation higher than wetland boundary
  Generally, soil moisture regime at these stations was the same as at Station C, except
    that it was progressively lower
Station E
  Located at the wetland limit
  Soil saturation at the 30 and 45 cm depths was observed for the latter half of the study
Station F
  Located within the wetland less than 0.5 m below wetland boundary elevation
  Soil moisture fluctuations did not differ from Stations D and E
  Soil saturation at the 30 cm depth was not observed
Station G
  Located within the wetland, approximately 0.75 m below wetland boundary elevation
  Soil saturation was observed at all depths throughout the study

REFERENCES

Carter, V., 1986. An overview of the hydrologic concerns related to wetlands in the United States. *Canadian Journal of Botany* 64:364–374.

Golet, F. C. and Lowry, D. J., 1987. Water regimes and tree growth in Rhode Island Atlantic white cedar swamps, in Laderman, A. D., ed., *Atlantic White Cedar Wetlands*. Westview Press, Boulder, CO.

Harris, R. W., 1983. *Arboriculture*. Prentice-Hall Inc., Englewood Cliffs, New Jersey. 688 p.

Ontario Ministry of Natural Resources, 1993. Ontario Wetland Evaluation System — Southern Manual. Ontario Ministry of Natural Resources, Toronto.

Pringle, E., 1995. Personal Communication. Agriculture and Agri-Food Canada. Guelph, Ontario.

Soil Equipment Corporation, 1987. Quick draw soilmoisture probe, 2900 F1. Technical Information. Santa Barbara, CA. 20 p.

Tiner, R., 1991. How wet is a wetland? *Great Lakes Wetlands* 2:1–7.

Topp, G. C., 1993. Soil water content, in Carter, M. R., ed., *Soil Sampling and Methods of Analysis*. Lewis Publishers for the Canadian Society of Soil Science, p. 541–549. CRC Press, Boca Raton, FL.

Westerman, R. L., 1990. *Soil Testing and Plant Analysis*. Third Edition. Number 3 in the Soil Science of America Book Series. Soil Science Society of America, Inc., Madison, WI.

7 ADAPTIVE MECHANISMS OF PLANTS OCCURRING IN WETLAND GRADIENTS

C. W. P. M. Blom, H. M. van de Steeg, and L. A. C. J. Voesenek

ABSTRACT

In river floodplains, vegetation zonation is strongly influenced by water-level fluctuations. Depending on elevation level and flood intensity, a natural as well as a human-influenced environmental gradient can be observed. The natural vegetation consists of softwood and hardwood forests. Grazed marsh and grassland are the characteristic communities of agriculturally used floodplains. Individual plants occurring in floodplains must develop adaptive mechanisms in order to survive transient floods. These mechanisms are discussed at the level of life history strategies and at the physiological level in relation to the position of the species in the floodplain gradient.

Species from the higher sites favor rapid germination after seed release while species from lower sites postpone germination or survive in the vegetative stage as a response to flooding. Flood-tolerant species are able to form aerenchymatous roots and survive submergence by their ability to accelerate shoot extension which restores leaf-air contact. Phytohormones, especially ethylene, play an important role in the adaptive responses to flooding.

The study of adaptive mechanisms of species occurring along the floodplain gradient will result in a better understanding of flooding-related processes acting at the community level.

INTRODUCTION

Wetlands constitute substantial areas of northwestern Europe, and in particular, The Netherlands, which in large part is situated below sea level. A system of dunes and dykes protects the land against flooding. In the past, human settlements were established mainly along rivers. In order to protect these settlements during

1-56670-147-3/96/$0.00+$.50

periods of flooding, the rivers were bounded by high, elevated dams — the so-called winter dykes. Most of these dykes were built between the twelfth and fourteenth centuries. Due to the nutrient-rich conditions of regularly inundated soils, lands adjacent to the rivers were cultivated for grazing cattle. To protect these grasslands in the floodplain against summer flooding, lower summer dykes were also built. In the past, peaks in precipitation in the catchment areas, as well as melting snow in the mountains of central Europe, caused flooding predominantly during winter and spring. However, in the last few decades, excessive precipitation has caused rivers to overflow more frequently during the summer periods. Improved upstream drainage and straightening meanders to facilitate shipping and to generate hydroelectric energy have resulted in water levels which exceed the heights of the summer dykes even during the growing season. A decline in agricultural use of the land is one of the consequences of the increase in the frequency of summer floods.

The ribbon-like floodplain zones characterized by winter and summer dykes are up to several kilometers wide and are all that is left of the former extensive delta system which constituted the geology and geomorphology of vast areas in the Netherlands. These floodplains strongly resemble the large natural river valleys which have largely disappeared in Europe and are quickly disappearing elsewhere (Maltby, 1991). Nowadays, conservationists are trying to protect the floodplains in northwestern Europe because of the growing awareness of the importance of river areas as nature reserves. Studies on natural plant and animal communities occurring in these areas are rapidly increasing. They are directed at nature management and also to find answers to fundamental scientific questions.

The background to the scientific explorations stems from the hypothesis that organisms occurring in areas that are exposed to irregular and unpredictable floods must develop adaptive mechanisms to overcome the adverse conditions of inundation.

This chapter attempts to investigate the diversity of plant communities in relation to the different conditions caused by transient flooding of river floodplains. The study also deals with the adaptive responses of individual plants to periods of submergence. Various life cycle strategies as well as adaptive responses at the morphological and physiological level are discussed in relation to the position of the species in the floodplain gradient.

REGIONAL SETTING AND DESCRIPTION OF STUDY SITE

The river Rhine originates in the Swiss Alps and passes through France, Germany, and the Netherlands. The length of this river is 1,250 km and the total drainage area is 185,000 km². The river Meuse rises in France and drains parts of France, Belgium, and the Netherlands. The length of the Meuse is 890 km and the total drainage area is about 33,000 km². In the Netherlands both rivers enter a lowland area, where they form a river delta before flowing into the North Sea. An essential difference between the rivers Rhine and Meuse is their water source: the Rhine is a combined glacier-rainwater river whereas the Meuse is fed by rainwater

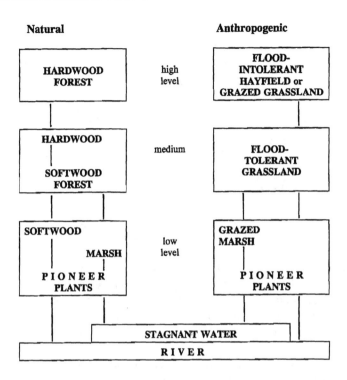

Figure 1 Natural and anthropogenic-influenced gradients in river floodplains of northwestern Europe.

only. Therefore the Rhine has a more stable discharge pattern over the seasons. The mean and maximum discharges of the Rhine at the Dutch border are 2,200 and 13,000 m^3 s^{-1}, respectively. The mean discharge of the Meuse is 250 m^3 s^{-1} and the maximum value is about 3,000 m^3 s^{-1} (Van den Brink *et al.*, 1993b).

EXPERIMENTAL METHODS

Both plant zonation and plant interactions are initiated by environmental gradients in floodplains (Van der Valk, 1981; Wilson and Keddy, 1986a; Shipley *et al.*, 1991). The important first step in understanding the relationships between environmental factors and zonation is to gain knowledge at the community level about species composition along gradients of riverine areas. Depending on the elevation level and human impact, a natural and an anthropogenic plant succession are described in floodplains of fluvial systems (Figure 1).

A useful way to understand the adaptive mechanisms of a plant community is to pursue experimental work. Field experiments at the community level have met with many of the problems recently emphasized by Gurevitch and Collins (1994). Therefore we chose to use species that are indicative of certain environmental conditions and could represent groups of plants that occur under similar conditions. In the present study, species from the genera *Rumex* (Blom *et al.*,

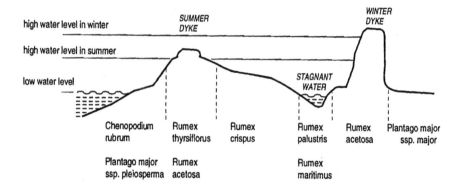

Figure 2   The distribution of *Rumex* species, *Plantago* species, and *Chenopodium rubrum* along wetland gradients of river areas in northwestern Europe.

1990; Laan, 1990; Voesenek, 1990), *Chenopodium* (Van der Sman, 1992), and *Plantago* (Blom, 1979; Lotz, 1989; Engelaar, 1994) which were identified as indicator species were used. They all occur in floodplains and their distribution depends on water level fluctuations and duration of the flooding periods, i.e., hydroperiods. Thus, they all can be found in the various zones typical of these areas. The distribution of these plants in different zones (gradients) is shown in Figure 2. Life cycle strategies of the indicator species are studied in relation to frequency and duration of the flooding periods. Adaptive mechanisms at the morphological and physiological level are investigated under laboratory conditions.

RESULTS

The Natural Gradient in and along the River (Table 1)

The vegetation zonation and species composition in and along a river (Table 1) is strongly influenced by water level fluctuations (Van Donselaar, 1961; Van den Brink *et al.*, 1993a; Admiraal *et al.*, 1993). In tidal areas, with regular fluctuations of fresh water, only two vegetation zones exist — a sublittoral zone with submerged and floating leaved aquatic vegetation, and a littoral zone dominated by *Phragmites australis*. Seasonal fluctuations with prolonged high water levels in spring eliminate *Phragmites australis* and leave a zone suitable for summer-germinating ephemerals.

Aquatic vegetation in the main channel is restricted to the tidal reaches of the river. The most common species is *Potamogeton pectinatus*. *P. perfoliatus* and *P. nodosus* occur in less fluctuating situations. On banks of the nontidal river the amphibious species *Polygonum amphibium* occurs. At high water levels this species is in its aquatic growth form, with long internodes and flexible leaves. At low water levels between two flooding periods the plants develop short, stiff, and hairy shoots. In frequently flooded former channels *P. amphibium* is joined by

Table 1   The Natural Vegetation Gradient in Floodplains

Vegetation type	Levee, river bank, and river on sand	Backswamp and connected former channel on clay	Backswamp and isolated former channel on clay
	High Level		
Hardwood forest	*Quercus robur* *Ulmus minor* Fraxinus excels. Acer campestre Acer pseudoplat.		
	Medium level		
Hardwood forest	*Ulmus minor* *Quercus robur* Fraxinus excels.	**Fraxinus excels.** Ulmus minor Quercus robur	**Fraxinus excels.** Ulmus minor Quercus robur
Softwood forest	*Salix alba* Populus nigra		
	Low level		
Softwood forest	*Salix alba* Populus nigra	Salix alba	
Marsh	*Phalaris arund.*	**Phalaris arund.** Carex acuta	**Phragmites austr.** *(Typha angustifol.)* *(Scirpus lacustris)*
Exposed bank	*Chenopodium rub.* C. glaucum Polygonum lapath.	**Plantago pleiosp.** Rumex palustris R. maritimus **Juncus bufonius** **Limosella aquat.**	**Rorippa amphibia** Oenanthe aquatica
Water	—	Polygonum amph. *(Nymphoides pelt.)*	**Nuphar lutea** Nymphaea alba *(Nymphoides pelt.)*

Note: Bold type   — dominant species.

Normal type — accompanying or locally abundant species.

( )   — severe damage in years with summer flooding, recovery very slow.

Nymphoides peltata. Former channels protected by summer dykes against summer flooding are mostly dominated by *Nuphar lutea* and more rarely by *Nymphaea alba* (Smits, 1994).

The littoral zone of rivers in the tidal area is dominated by *Phragmites australis*. On the higher parts of the river bank *P. australis* is accompanied by *Calystegia sepium*. Upstream of the tidal reach *P. australis* disappears rather suddenly. *Phalaris arundinacea* replaces *P. australis* on banks of the nontidal river. On banks of former channels with sufficient protection against summer floods the number of marshland species strongly increases. *Phragmites australis*, *Typha angustifolia, Scirpus lacustris* and *Glyceria maxima* are entirely restricted to former channels without summer flooding. High summer floods have eliminated *Sparganium erectum* and *Ranunculus lingua* and decimated *Typha angustifolia* and *Scirpus lacustris* on the banks of this former channel type (Van de Steeg, 1984, Brock *et al.,* 1987).

Characteristic of the rather dry river bank are summer annuals like *Chenopodium rubrum, C. glaucum,* and *Polygonum lapathifolium.* Mudflat species occur only occasionally in vegetation types dominated by *Chenopodium* species and *P. lapathifolium.*

The species composition of mudflats is regulated by the effects of specific flooding conditions on germination, growth, and seed production of individual species and by the effects of frequency of exposure to dry conditions occurring between two floods (Voesenek and Blom, 1992). Mudflats along former channels are colonized soon after exposure by short-lived species, ranging from a rather dry zone of *Plantago major* ssp. *pleiosperma,* via an intermediate zone of *Juncus bufonius* with *Rumex palustris* and *R. maritimus,* to a permanent wet zone of *Limosella aquatica.*

Late exposure and permanent wet conditions in mudflats of more isolated former channels result in abundant germination of ephemeral marsh species such as *Rorippa amphibia* and *Oenanthe aquatica.*

The natural vegetation of floodplains in temperate Western and Middle Europe consists of softwood species in the youngest parts of the floodplain and hardwood species in the older parts (Table 1). The lower sites of floodplains along Alpine rivers with a summer flooding regime and with coarse substrates are dominated by *Alnus incana.* With the ongoing siltations of floodplains, the flood-resistant pioneer forest of *Alnus incana* is replaced by a mesophilic forest type of *Fraxinus excelsior* and *Ulmus* species. Pioneer species on the banks of rivers with a pluvial discharge regime are *Salix alba* and *Populus nigra. Salix alba* is the most common softwood species along European rivers of this discharge type. In drier situations *P. nigra* dominates the river banks. *Salix alba* is the characteristic softwood forest species of the Danube in Central Hungary (Kárpáti and Tóth, 1961), of the Upper Rhine in Germany (Dister, 1980), and of the Lower Rhine in the Netherlands. *Populus nigra* is the dominant softwood forest species of river banks along the Loire in France. Without interference by man, the pioneer forest types of both softwood species will gradually be replaced by a hardwood forest consisting of *Quercus robur, Ulmus laevis, U. minor, Fraxinus excelsior,* or *F. angustifolia* (Kárpáti and Tóth, 1961; Dister, 1980; Carbiener and Schnitzler, 1990; Schnitzler *et al.,* 1992).

The Anthropogenic Vegetation Gradient in Floodplains (Table 2)

The species composition of floodplain grasslands (Table 2) is dependent on flooding pressure, soil conditions, and management practices (Sykora *et al.,* 1988; Jongman, 1992). Grassland is mostly confined to the high and medium levels of floodplains. On lower levels of the river bank, trampling by grazing animals results in the destruction of *Phalaris arundinacea* and the invasion of the grazing-resistant *Cirsium arvense.* On permanent wet banks of former channels *Phragmites australis,* which is a trampling-sensitive species, is replaced by the trampling-resistant *Glyceria maxima.* In the less flooded areas, land use practices such as haymaking and grazing result in distinct differences in vegetation height, species

Table 2   Zonation of Anthropogenic Floodplain Vegetation

Floodplain level/management	Geomorphological elements substrate, and main flooding season	
	Riverbank and river, sand, winter and spring flooding	Levee, backswamp, and former channel, clay, winter flooding
High		
Grazing	*Festuca rubra*	*Lolium perenne*
	Medicago falcata	*Trifolium repens*
	Ranunculus bulbosus	*Ranunculus acris*
	Rumex thyrsiflorus	Rumex acetosa
	Salvia pratensis	*Cirsium arvense*
	Eryngium campestre	*Bellis perennis*
Haymaking	*Festuca rubra*	*Arrhenatherum elatius*
	Avena pubescens	*Trisetum flavescens*
	Medicago falcata	*Trifolium pratense*
	Ranunculus bulbosus	*Ranunculus acris*
	Rumex thyrsiflorus	*Rumex acetosa*
	Salvia pratensis	*Plantago lanceolata*
		Crepis biennis
		Peucedanum carvifolia
Medium		
Grazing and haymaking		
	Elymus repens	*Lolium perenne*
	Potentilla reptans	*Poa trivialis*
	Inula britannica	*Ranunculus repens*
	Festuca arundinacea	Alopecurus pratensis
	Rumex crispus	Rumex crispus
	(Rorippa sylvestris)	*(Rorippa sylvestris)*
	(Agrostis stolonif.)	*(Agrostis stolonif.)*
Low		
Grazed	*Cirsium arvense*	*Glyceria maxima*
Ungrazed	*Phalaris arundinacea*	*Phragmites australis*
		Typha angustifolia
		Scirpus lacustris

Note: Bold type   — species dominant or abundant.
Normal type — accompanying species or locally abundant.
(   )   — dominant species after summer flooding.

composition, and species number. Mowing leads to a much higher species richness with *Arrhenatherum elatius, Trisetum flavescens, Crepis biennis, Tragopogon pratense, Peucedanum carvifolia,* and *Trifolium pratense.* In grazed grasslands, only grazing- and trampling-resistant species like *Lolium perenne, Poa trivialis, Bellis perennis, Trifolium repens,* and *Taraxacum officinale* are able to survive. A species that takes considerable advantage of grazing floodplain grasslands is *Cirsium arvense.* Management practices have minor impact on the species composition of frequently flooded medium-level floodplain grassland.

Medium-level floodplain grasslands on sandy soil are dominated by *Elymus repens* and *Potentilla reptans. Inula britannica, Festuca arundinacea,* and *Achillea ptarmica* are almost always restricted to this grassland type. Frequently flooded grasslands on clay soils are dominated by *Agrostis stolonifera, Poa trivialis,* and *Ranunculus repens. Rumex crispus* is characteristic of both flood-tolerant grassland types.

At higher elevations, distinct changes in species composition take place. Flood-tolerant species become far less abundant and a large number of new species occurs. Common species of less flooded grassland types are *Rumex acetosa, Lolium perenne, Plantago lanceolata, Bellis perennis,* and *Ranunculus acris.*

In hayfields *Arrhenatherum elatius* and *Trisetum flavescens* become dominant. On sandy parts of the levees the aforementioned mesophilic species lose their dominance, leaving space for species with adaptations to dry soil conditions such as *Festuca rubra, Ranunculus bulbosus, Eryngium campestre, Medicago falcata, Salvia pratensis,* and *Rumex thyrsiflorus.*

In years with extremely high water levels in summer, all floodplains will be flooded. During a flooding period of only a few days, damage of the vegetation on high floodplain levels is restricted. In the medium-level floodplain flooding will last several weeks. In the very turbid and rather warm water the entire vegetation dies back. After recession of the water some species regenerate from tap roots (*Rumex crispus*) or rhizomes (*Elymus repens*). Most species, however, have to regenerate from seed. Species least disadvantaged by catastrophic summer flooding are grassland species with persistent seedbanks like *Poa trivialis* and *Rumex crispus.* In persistent wet situations *Rorippa sylvestris* will become dominant.

Life Cycle Strategies

Hypotheses for studies on life-history strategies are derived from field observations of some plant species that are able to survive the extreme submergence conditions by adaptations at different life cycle levels. The success of plants to survive flooding strongly depends on the interaction between the stage in the life cycle and both the timing of the start and duration of the flooding period. Important stages of the life cycle are the seed stage, germination, seedling establishment, growth, vegetative spread, flowering, seed production, and seed dispersal. Among the three typical plant life histories — annuals, biennials, and perennials — the annuals have certain advantages in constantly disturbed environments (cf. Grime, 1979). Annuals and biennials, especially, grow together in the same temporary habitats. Van der Meijden *et al.,* (1992) showed that due to their dispersal characteristics, biennials are particularly suited for exploiting temporary habitats.

To study the phenomenon of adaptive strategies at the life cycle level in detail, choice was made of the "one species — one habitat" approach in which various species growing under the same conditions in the field were compared (Van der Sman, 1992). Based on that approach, *Rumex maritimus, R. palustris,* and *Chenopodium rubrum* were chosen to investigate the growth characteristics of different species, all occurring in the low areas of floodplains. These species have to cope with the adverse conditions generated by the intense floods. These species differ in their adaptation towards flooding and this difference is related to their life histories. *Rumex maritimus* is an annual or a biennial, *R. palustris* is

Table 3   Percentage Flowering Per Cohort and Treatment in *Chenopodium rubrum, Rumex maritimus,* and *Rumex palustris*

Cohort	C. rubrum		R. maritimus		R. palustris	
	Drained	Flooded	Drained	Flooded	Drained	Flooded
May	100	0	100	88	95	0
June	100	100	100	35	0	0
August	100	100	0	0	0	0
September	45	21	0	0	0	0

Note: Plants were sown 32 days before the start of a 10-day flooding period. May cohorts were flooded 4 times with 6-week intervals. June, August and September cohorts were flooded 3, 2, and 1 time, respectively.

From Van der Sman, A. J. M. *et al., Journal of Ecology,* 81:121–130, 1993. With permission.

predominantly a biennial, and *Chenopodium rubrum* is strictly an annual. Flood tolerance of these species, in relation to their developmental stage, was studied in the field as well as in large-scale outdoor experiments. Although all three species occupy the same habitat, they differ markedly in adaptive strategies. Survival during submergence in the prereproductive phase was high in both *Rumex* species, but low in *C. rubrum* (Van der Sman et al., 1993a). Remarkable variation was observed in flowering characteristics between early and late emerging plants of *C. rubrum* (Table 3). Flooding scarcely reduced the probability of flowering in *C. rubrum,* although submerged plants hardly managed to produce seeds. However, plants of the same cohort which emerged later in the season were able to start flowering at an early stage and completed their life cycle in a very short time. This characteristic enables populations of *C. rubrum* to react very quickly in the short dry periods between two floods. Different results were observed in both species of *Rumex* (Van der Sman et al., 1993a). Flowering stems were initiated only when a minimum number of leaves were present and cohorts raised later remain vegetative. Flooding delayed flowering until later in the season or even until the following year. *Rumex maritimus* flowered more rapidly and was more susceptible to flooding than *R. palustris.* In *R. maritimus,* survival following flooding was restricted to plants which emerged from the water surface by elongating their main shoot (Van der Sman et al., 1993b). In early summer, plants of *R. palustris* that were flooded and remained below the water line were better able to survive; flowering was delayed or even postponed to the following season (Table 3).

To understand the factors that affect species distribution along gradients, it is important to know the survival chances of the plants during the early stages of their life cycle in relation to fluctuating conditions in the environment. The stages of germination and establishment of seedlings are probably the most vulnerable periods in the life cycle of plants (Cavers and Harper, 1967; Marks and Prince, 1981; Fenner, 1987). In regularly disturbed and harsh environments both late and multiple germination will benefit survival (Andersson, 1990; Badger and Ungar, 1991). The type of seedbank, transient or persistent (Thompson and Grime, 1979), dormancy status (Karssen, 1982), and germination requirements (Washitani and Masuda, 1990) determine timing of germination. A comparison was made of the field germination characteristics and seedbank features of *Rumex acetosa,*

R. crispus, and *R. palustris,* which occur at different zones in the floodplain (Figure 2). In reciprocal transplant experiments in the floodplain, emergence of the three species was studied in dry, intermediate, and wet zones, which represent the habitats of the species under study (Voesenek and Blom, 1992).

In *R. acetosa,* early seed release and easy separation of the perianth and achene facilitate fast germination in the autumn. This species possesses a transient seedbank. A small amount of still-ungerminated achenes enters the soil in autumn and will then germinate in the following spring or summer. The reciprocal approach proved that waterlogging of the soil prevents germination of *R. acetosa* (Figure 3). The high mortality of the seeds demonstrated the vulnerability to flooding. Germination of *R. crispus* and *R. palustris* immediately after achene drop is very limited and germination occurs in the next spring. This delay can be partly due to the inhibition of the germination process by the presence of a perianth (Cavers and Harper, 1967; Voesenek and Blom, 1992). Both *R. crispus* and *R. palustris* showed considerable germination in the *crispus* habitat compared to the *acetosa* zone. Flooding also caused secondary dispersal, a second germination flush, and burial in the seedbank. Results of seedbank experiments showed that these species have a persistent seedbank (Voesenek and Blom, 1992). Percentage of emergence of *R. crispus* corresponds positively with increasing soil moisture levels. New seedlings of both *R. crispus* and *R. palustris* emerged in the *crispus* zone immediately after a summer flood but not after an autumn flood.

Adaptive Responses of Below-Ground Organs

After flooding or soil waterlogging, first reactions occur in the soil. The oxygen available for plant roots and microorganisms rapidly decreases as the soil water level rises (Laan *et al.,* 1989b). At least two changes can be observed in soils after flooding. First is the development of aerenchyma. Plants require oxygen for cell division and when the gas exchange between the atmosphere and the soil is hampered roots of many species are affected, at least during periods of growth involving cell division. As a response to these conditions, flood-resistant species are able to generate a new root system. These roots are characterized by aerenchyma formed either by cell differentiation and cell collapse — lysigenous aerenchyma — or by cell separation without collapse, defined as schizogenous aerenchyma (Jackson and Drew, 1984; Justin and Armstrong, 1987; Crawford, 1992). Well-developed aerenchyma — which is a charateristic feature of above- and below-ground parts of many wetland and aquatic species — provides large and continuous air spaces that facilitate the downward diffusion of oxygen from shoot to root (Armstrong and Armstrong, 1988; Laan *et al.,* 1989a).

A second effect of flooding is the inhibition of microbiological activities. In organic-rich waterlogged soils, hypoxic conditions are established by decreased oxygen diffusion and by oxygen consumption of soil microorganisms. Nitrification is one of the oxygen-dependent processes that will be inhibited by soil flooding.

An examination was made of the root development of the indicator species of *Rumex* after waterlogging (Laan *et al.,* 1989a). Under experimental conditions

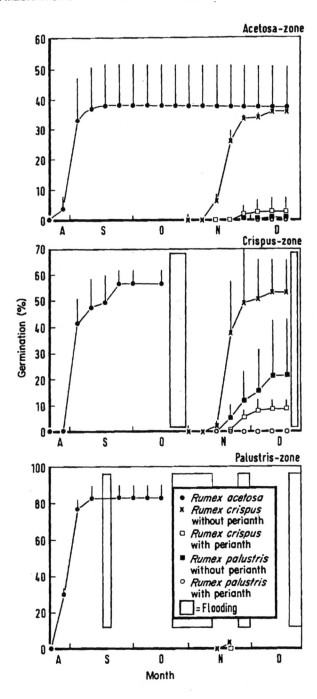

Figure 3 Germination characteristics of three *Rumex* species in three different zones (see Figure 2) in the river floodplain. (After Voesenek, L. A. C. J. and Blom, C. W. P. M., *Acta Botanica Neerlandica*, 41:319–329, 1992. With permission.)

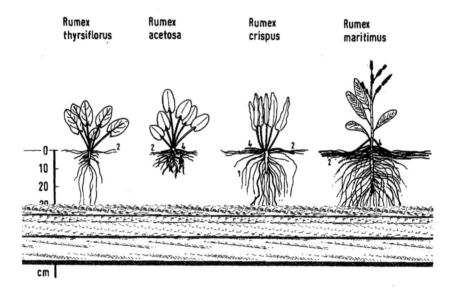

Figure 4   Root development of *Rumex* species after three weeks of soil flooding. Root type: 1 and 3, downward-growing laterals; 2, horizontally growing laterals; 4, adventitious roots. (After Laan, P. *et al., Journal of Ecology*, 77:693–703, 1989. With permission.)

and after two weeks of soil flooding, the total length of all new roots was found to be much higher in the *Rumex* species originating from low-lying and overwet sites in the river area than in *Rumex* plants from the drier habitats (Figure 4). In the species of low-lying areas more than 20% of the old root system was replaced by new roots. Almost no new laterals were observed in the species at higher levels. Development and outgrowth of new laterals were well correlated with the capability of a species to form aerenchyma. The high porosity in *R. maritimus* roots could largely explain the observation that these roots keep growing when flooded. In the high floodplain species *R. thyrsiflorus*, porosity values of the roots were generally lower than 10%, whereas *R. maritimus* and some other species from wet habitats showed values higher than 10%. These results were in accordance with the findings of Smirnoff and Crawford (1983) who distinguished between tolerant and intolerant species at 10% root porosity. The development of the new porous roots mainly occurred at the water surface and new roots grew horizontally as well as downwards (Figure 4). The growth rate of roots and shoots depended on both the number and porosity of the roots. The ability to form new fast-growing roots was apparently the result of aerenchyma formation. Root porosity of species from the drier sites did not increase. *Rumex maritimus*, especially, was able to quickly develop new aerenchymatous roots, which may help to explain its location in the field. This species predominantly occurs at lower-lying sites with highly fluctuating water levels.

Aerenchyma not only enables plant roots to grow under nearly anaerobic soil conditions but, due to a surplus of oxygen transported to the root tips, oxygen leakage through the roots could maintain oxidized conditions in a small zone around the individual roots (Armstrong and Armstrong, 1988). Therefore a positive effect on nitrification can be expected in the rhizosphere due to radial oxygen loss. Engelaar *et al.* (1991, 1993a) compared the effects of radial oxygen loss of the flooding-resistant *R. palustris* with that of the flooding-intolerant *R. acetosa* on nitrification. Due to a positive effect of radial oxygen loss by *R. palustris* on nitrification, growth of these plants under waterlogged conditions was hardly inhibited. In contrast, the rhizosphere of *R. acetosa* remained anaerobic because of the species' inability to form aerenchyma. Growth of this species was severely inhibited by soil flooding.

However, due to the weakness of porous roots *R. palustris* was not able to survive the soil compaction which can occur between two flooding periods. Grazing in the floodplains during dry periods, either for commercial purpose or for nature management, is a common land use practice. Settling of soil particles as a result of the withdrawal of water or as a consequence of trampling by cattle strongly increases the mechanical resistance of the soil. In an experimental study, the reactions of three species, all occurring in the river area but differing in field distribution and ability to form aerenchyma, were compared in their ability to resist flooding and compaction (Engelaar *et al.*, 1993b). Hypoxia due to flooding resulted in higher root porosities of *R. palustris* but not in *Plantago major*. With respect to soil compaction, *P. major* was the least affected species, followed by *R. acetosa* and *R. palustris* (Figure 5).

Adaptive Responses in Above-Ground Organs

A remarkable response to flooding in flood-resistant plants is enhanced shoot elongation. Accelerated growth enables submerged plants to restore contact with the open air. Exchange of atmospheric gases, mainly through the stomata, then allows the plant to continue growth. Many aquatic and amphibious and some terrestrial plants are able to enhance the elongation rates of petioles and/or internodes in order to emerge above the water surface (e.g., Kozlowski, 1984; Crawford, 1987; Blom, 1990; Blom *et al.*, 1990, 1993, 1994; Voesenek, 1990, Voesenek and Van der Veen, 1994). A close relationship between the ability to elongate shoots under water and aerenchyma development can be expected (Jackson, 1990; Voesenek *et al.*, 1992; Crawford, 1992). Moreover, the adaptive importance in achieving as large a leaf area as possible above the water surface is unquestionable (Voesenek and Blom, 1989; Laan *et al.*, 1990). An excellent example of the functional relationship between the emergence out of the water, shoot elongation, and above-water biomass can be found in the work of Van der Sman *et al.* (1991). Seed production of *Rumex maritimus* was positively correlated with shoot height above the water (Table 4). Laan *et al.* (1990) found that the amount of internal aeration in *R. maritimus* was positively correlated with the leaf area above the water surface. Van der Sman *et al.* (1991) proved that plants of this species, if submerged prior to flowering, were able to elongate under water. Moreover, if plants were submerged during the first stages of rosette growth, they

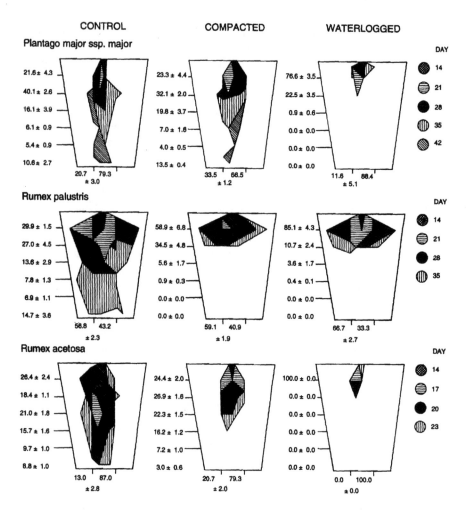

Figure 5   Outline of the root systems of three plant species in three soil types. Percentages of horizontal and vertical roots are indicated. (From Engelaar, W. M. H. G. *et al.*, *Acta Botanica Neerlandica*, 42:25–35, 1993. With permission.)

were better able to extend a larger part of the shoot above water and produce more seeds than those submerged later.

Both petiole and stem elongation in *Rumex* species can be correlated with increased internal concentrations of the gaseous plant hormone, ethylene (Voesenek and Blom, 1989; Voesenek et al., 1990, 1993a,b; Van der Sman et al., 1991). It is already well known that the gaseous growth substance, ethylene, promotes not only shoot elongation but also adventitious rooting and aerenchyma formation (Ku et al., 1970; Musgrave et al., 1972; Osborne, 1984; Jackson, 1985, 1990; Jackson et al., 1985, 1987; Yamamoto and Kozlowski, 1987a,b; Ridge, 1987; Laan and Blom, 1990; Schwegler and Brändle, 1991; Crawford, 1992). A new

Table 4   Correlation Between Stem
Length and Seed Production

Length class of the main stem (cm)	Treatment	
	Drained	Flooded
<50	18.6 ± 1.5	3.7 ± 0.2
50–54	18.3 ± 1.2	7.3 ± 0.4
>54	19.2 ± 0.2	11.4 ± 1.0

Note: Mean seed production in g/plant (±1 SE) per stem length in *Rumex maritimus*. Plants from seeds germinated in May were flooded for 4 weeks in July.

From Van der Sman, *et al.*, *Functional Ecology*, 5:304–313, 1991.

ecophysiological approach is to compare the relationship between ethylene production and adaptive responses of various species within one genus distributed along a wetland gradient. Voesenek and Blom (1989) demonstrated that the stimulated growth of petioles in two wetland species of *Rumex* after submergence can be simulated by exposing nonsubmerged plants of these species to ethylene. *Rumex* species normally occurring under dry conditions did not react to C_2H_4 application. The fact that application of an inhibitor of ethylene action ($AgNO_3$) inhibits shoot elongation and that the endogenous ethylene concentration increases during submergence strongly suggest that this phytohormone plays a regulatory role in the growth responses of these *Rumex* species under submerged conditions. Recent observations (Voesenek et al., 1993b) proved that both *R. acetosa*, a species occurring in dry habitats, and *R. palustris*, a representative of species highly adapted to flooding, accumulate ethylene under water. The first species shows no stimulation of leaf elongation whereas the second one significantly enhances the growth rate of underwater shoots. The behavior of *R. palustris* reflects a greater responsiveness than *R. acetosa* towards ethylene. These different reactions are in accordance with their different positions in the floodplain gradient.

DISCUSSION AND CONCLUSIONS

Life cycle strategies developed in plants to survive the stressful conditions during periods of inundation are often dependent on a rapid transition from an active life to a more passive "wait and see" behavior. Part of that strategy can involve the ability to flower and to produce seed immediately after conditions improve.

Reproductive habits, as an aspect of the life history traits of potential colonizers, strongly influence the success of colonization of floodplains. In an Alaskan floodplain, Walker et al. (1986) demonstrated that both life history traits and stochastic events related to flooding and seed dispersal determine primary succession.

The different germination characteristics and emergence patterns between *R. acetosa* on the one hand, and *R. crispus* and *R. palustris* on the other, as well as their different types of seedbanks, can at least partly explain the position of these species along the wetland gradient in the riverine floodplains. The species from the higher sites favor a transient seedbank and rapid germination after seed release. Summer mowing, grazing, and the activities of moles result in grasslands with an open structure and with gaps in which early emergence of *R. acetosa* can occur. This strategy contributes to a high competitive performance in the following growing season (see also Van der Toorn and Pons, 1988) and in these habitats survival chances do not decrease because of the absence of disturbance by flooding. Germination in species from the low-lying and flooded sites is delayed until the next growing season. *Rumex crispus* and *R. palustris* occur on sites in the river floodplain with predictable and extremely long winter floods. Germination just before the winter period would probably result in a very high seedling mortality and thus in a decreased fitness. The present study clearly shows a life history strategy based on timing of the germination and directed to an increase in chances for survival. The investigations of Baskin and Baskin (1972), Arthur *et al.* (1973), Marks and Prince (1981), and Venable (1984) have also demonstrated a similar type of life history strategy. Early germination favors reproduction, whereas late germination favors survival of the seedlings. After germination, many factors may influence chances of establishment. Interference between neighboring species (e.g., Wilson and Keddy, 1986b), light conditions (Menges and Waller, 1983), soil nutrient status (e.g., Shipley and Keddy, 1988; Campbell *et al.*, 1991; Campbell and Grime, 1992), and the level of plasticity these species have in order to cope with these changing environmental conditions upon flooding (Wilson, 1991) determine the success of their establishment in river floodplains.

In order to complete their life cycle, plants have to be adapted to the adverse conditions in flooded habitats, not only during the generative but also during the vegetative phase.

Morphology as well as physiology may change in response to flooding. A high degree of physiological plasticity as a response to inundation due to floods will be reflected in the rapid changes of root and shoot morphology. The number, rate of formation, and porosity of new roots can be associated with the elevational distribution of indicator species in floodplain gradients (Laan *et al.*, 1989a). As a reaction to flooding, species of low-lying areas formed more new roots and did so faster than species from higher regions. Due to aerenchyma formation, they also have higher porosities which have consequences for radial oxygen loss. The oxygen concentration in the rhizosphere of aerenchymatous plants is quickly restored upon flooding. This resulted in renewed growth and activities of ammonium- and nitrite-oxidizing bacteria and thus in the restoration of the nitrification process (Engelaar *et al.*, 1991). A drawback of a porous root system is the rapid collapse of the roots after soil compaction which occurs as a result of trampling or settling of soil particles by the withdrawal of water.

A strong relationship between aerenchyma development and above-ground adaptations can also be expected. Resistant plants survive submergence by their ability to accelerate shoot extension, which brings the upper leaves back into contact with the aerial environment. Other wetland species, such as *Ranunculus sceleratus* (Ridge, 1987) also respond with enhanced elongation rates as an opportunistic strategy to overcome submergence.

Using survival and seed production to determine fitness, Van der Sman *et al.* (1991) demonstrated increasing survival chances due to the positive relationship between shoot height above the water and seed production.

Hormones play an important role in the adaptive responses to flooding of plants occurring in wetland gradients. The gaseous plant hormone, ethylene, plays a crucial role in the signal-transduction pathway leading to stimulated shoot growth. To regulate this "depth accommodation" this growth hormone acts in concert with other plant hormones and the gases, oxygen and carbon dioxide (Jackson, 1990; Blom *et al.*, 1994; Voesenek and Van der Veen, 1994). This regulation is complex and far from being completely understood. However, in the genus *Rumex*, the degree of sensitivity to ethylene remarkably corresponds with the location of the different species along the gradient in floodplains. Species poorly adapted to flooding scarcely react even at considerably high concentrations of ethylene, whereas flooding-adapted species possess a high responsiveness to this hormone.

In conclusion, the present research on adaptive mechanisms of plants occurring in floodplain gradients proves that the use of indicator species, each representing particular communities, enables elucidation of processes directed to survival responses under flood conditions. The study of more species belonging to one genus proved to be a useful approach to get an insight into the adaptive mechanisms aimed at resisting adverse environmental conditions. Comparing reactions of these related species with the behavior of plants from other genera which occur in the same habitats will certainly result in a better understanding of the processes acting at the community level.

REFERENCES

Admiraal, W., Van der Velde, G., Smit, H., and Cazemier, W. G., 1993. The rivers Rhine and Meuse in The Netherlands: present state and signs of ecological recovery. *Hydrobiologia* 265:97–128.

Andersson, S., 1990. The relationship between seed dormancy, seed size and weediness, in *Crepis tectorum* (Asteraceae). *Oecologia* 83:277–280.

Armstrong, J. and Armstrong, W., 1988. *Phragmites australis* — A preliminary study of soil-oxydizing sites and internal gas transport pathways. *New Phytolologist* 108:373–382.

Arthur, A. E., Gale, J. S., and Lawrence, K. J., 1973. Variation in wild populations of *Papaver dubium*. VII. Germination time. *Hereditas* 30:189–197.

Badger, K. S. and Ungar, I. A., 1991. Life history and population dynamics of *Hordeum jubatum* along a soil salinity gradient. *Canadian Journal of Botany* 69:384–393.

Baskin, J. M. and Baskin, C. C., 1972. The influence of germination date on survival and seed production of *Leavenworthia stylosa*. *American Midland Naturalist* 88:318–323.

Blom, C. W. P. M., 1979. Effects of Trampling and Soil Compaction on the Occurrence of Some *Plantago* Species in Coastal Sand Dunes. Ph.D. Thesis, University of Nijmegen, 135 pp.

Blom, C. W. P. M., 1990. Responses to flooding in weeds from river areas, in Kawano, S., ed., *Biological Approaches and Evolutionary Trends in Plants*. London: Academic Press, p. 81–94.

Blom, C. W. P. M., Bögemann, G. M., Laan, P., Van der Sman, A. J. M., Van de Steeg, H. M., and Voesenek, L. A. C. J., 1990. Adaptations to flooding in plants from river areas. *Aquatic Botany* 38:29–47.

Blom, C. W. P. M., Voesenek, L. A. C. J., and Van der Sman, A. J. M., 1993. Responses to total submergence in tolerant and intolerant riverside species, in Jackson, M. B. and Black, C. R., eds., *Interacting Stresses on Plants in a Changing Climate*, Springer-Verlag, Berlin, pp. 243–266.

Blom, C. W. P. M., Voesenek, L. A. C. J., Banga, M., Engelaar, W. M. H. G., Rijnders, J. H. G. M., van de Steeg, H. M., and Visser, E. J. W., 1994. Physiological ecology of riverside species, adaptive responses upon submergence. *Annals of Botany* 74:253–263.

Brock, Th.C. M., Velde, G. van de Steeg, H. M., 1987. The effects of extreme water level fluctuations on the wetland vegetation of a nymphaeid dominated oxbow lake in The Netherlands. *Archiv für Hydrobiologie, Ergebnisse der Limnologie* 27:57–73.

Campbell, B. D., Grime, J. P., and MacKey, J. M. L., 1991. A trade-off between scale and precision in resource foraging. *Oecologia* 87:532–538.

Campbell, B. D. and Grime, J. P., 1992. An experimental test of plant strategy theory. *Ecology* 73:15–29.

Carbiener, R. and Schnitzler, A., 1990. Evolution of major pattern models and processes of alluvial forest of the Rhine in the rift valley (France/Germany). *Vegetatio* 88:115–129.

Cavers, P. B. and Harper, J. L., 1967. Studies on the dynamics of plant populations. I. The fate of seed and transplants introduced into various habitats. *Journal of Ecology* 55:59–71.

Crawford, R. M. M., 1987. *Plant Life in Aquatic and Amphibious Habitats*, Blackwell Scientific, Oxford.

Crawford, R. M. M., 1992. Oxygen availability as an ecological limit to plant distribution. *Advances in Ecological Research* 23:93–185.

Dister, E., 1980. Geobotanische Untersuchungen in der Hessischen Rheinaue als Grundlage für die Naturschutzarbeit. Ph.D. Thesis, University of Göttingen, 170 pp.

Engelaar, W. M. H. G., 1994. Roots, Nitrification and Nitrate Acquisition in Waterlogged and Compacted Soils. Ph.D. Thesis, University of Nijmegen, 175. pp.

Engelaar, W. M. H. G., Bodelier, P. L. E., Laanbroek, H. J., and Blom, C. W. P. M., 1991. Nitrification in the rhizosphere of a flooding-resistant and a flooding-non-resistant *Rumex* species. *FEMS Microbiology and Ecology* 86:33–42.

Engelaar, W. M. H. G., van Bruggen, M. W., van den Hoek, W. P. M., Huyser, M. A. H., and Blom, C. W. P. M., 1993a. Root porosities and radial oxygen losses of *Rumex* and *Plantago* species as influenced by soil-pore diameter and soil aeration. *New Phytologist* 125:565–574.

Engelaar, W. M. H. G., Jacobs, M. H. H. E., and Blom, C. W. P. M., 1993b. Root growth of *Rumex* and *Plantago* species in compacted and waterlogged soils. *Acta Botanica Neerlandica* 42:25–35.

Fenner, M., 1987. Seedlings. *New Phytologist* 106 (suppl.):35–47.

Grime, J. P., 1979. *Plant Strategies and Vegetation Processes.* John Wiley & Sons, New York.

Gurevitch, J. and Collins, S. L., 1994. Experimental manipulation of natural plant communities. *Trends in Ecological Evolution* 9:94–98.

Jackson, M. B., 1985. Ethylene and responses of plants to soil waterlogging and submergence. *Annual Review in Plant Physiology* 36:145–174.

Jackson, M. B., 1990. Hormones and developmental change in plants subjected to submergence or soil waterlogging. *Aquatic Botany* 38:49–72.

Jackson, M. B., Davies, D. D. and Lambers, H., eds., 1991. *Plant Life under Oxygen Deprivation. Ecology, Physiology and Biochemistry.* SPB Academic Publishing, The Hague, pp. 3–21.

Jackson, M. B. and Drew, M. C., 1984. Effects of flooding on growth and metabolism of herbaceous plants, in Kozlowski, T. T., ed., *Flooding and Plant Growth*, Academic Press, Orlando, FL, pp. 47–128.

Jackson, M. B., Fenning, T. M., Drew, M. C., and Saker, L. C., 1985. Stimulation of ethylene production and gas space (aerenchyma) formation in adventitious roots of *Zea mays* L. by small partial pressures of oxygen. *Planta* 165:486–492.

Jackson, M. B., Waters, I., Setter, T., and Greenway, H., 1987. Injury to rice plants caused by complete submergence; a contribution by ethylene (ethene). *Journal of Experimental Botany* 38:1826–1838.

Jongman, R. H. G., 1992. Vegetation, river management and land use in the Dutch Rhine floodplains. *Regulated Rivers* 7:279–289.

Justin, S. H. F. W. and Armstrong, W., 1987. The anatomical characteristics of roots and plant response to soil flooding. *New Phytologist* 106:465–495.

Kárpáti, I. and Tóth, I., 1961. Die Auenwaldtypen Ungarns. *Acta Agronomica Academie Scientiarum Hungaricorum* 11:421–452.

Karssen, C. M., 1982. Seasonal patterns of dormancy in weed seeds, in Khan, A. A., ed., *The Physiology and Biochemistry of Seed Development, Dormancy and Germination*, Elsevier Biomedical Press, Amsterdam, pp. 243–270.

Kozlowski, T. T., 1984. *Flooding and Plant Growth.* Academic Press, London.

Ku, H. S., Suge, H., Rappaport, L., and Pratt, H. K., 1970. Stimulation of rice coleoptile growth by ethylene. *Planta* 90:333–339.

Laan, P., 1990. Mechanisms of Flood-Tolerance in *Rumex* Species. Ph.D. Thesis, University of Nijmegen, 159 pp.

Laan, P., Berrevoets, M. J., Lythe, S., Armstrong, W., and Blom, C. W. P. M., 1989a. Root morphology and aerenchyma formation as indicators of the flood-tolerance of *Rumex* species. *Journal of Ecology* 77:693–703.

Laan, P., Smolders, A., Blom, C. W. P. M., and Armstrong, W., 1989b. The relative roles of internal aeration, radial oxygen losses, iron exclusion and nutrient balances in flood-tolerance of *Rumex* species. *Acta Botanica Neerlandica* 38 131–145.

Laan, P. and Blom, C. W. P. M., 1990. Growth and survival responses of *Rumex* species to flooded and submerged conditions: The importance of shoot elongation, under water photosynthesis and reserve carbohydrates. *Journal of Experimental Botany* 41:775–783.

Laan, P., Tosserams, M., Blom, C. W. P. M., and Veen, B. W., 1990. Internal oxygen transport in *Rumex* species and its significance for respiration under hypoxic conditions. *Plant and Soil* 122:39–46.

Lotz, L. A. P., 1989. Variation in Life-History Characteristics Between and Within Populations of *Plantago major* L. Ph.D. Thesis, University of Nijmegen, 175 pp..

Maltby, E., 1991. Wetlands — their status and role in the biosphere, in Jackson, M. B., Davies, D. D., and Lambers, H., eds., *Plant Life under Oxygen Deprivation. Ecology, Physiology and Biochemistry*, SPB Academic Publishing, The Hague, pp. 3–21.

Marks, M. and Prince, S., 1981. Influence of germination date on survival and fecundity in wild lettuce *Lactuca serriola. Oikos* 36:326–330.

Menges, E. S. and Waller, D. M., 1983. Plant strategies in relation to elevation and light in floodplain herbs. *American Naturalist* 122:454–473.

Musgrave, A., Jackson, M. B., and Ling, E., 1972. *Callitriche* stem elongation is controlled by ethylene and gibberellin. *Nature* 238:93–96.

Osborne, D. J., 1984. Ethylene and plants of aquatic and semi-aquatic environments. A review. *Plant Growth Regulators* 2:167–185.

Ridge, I., 1987. Ethylene and growth control in amphibious plants, in Crawford, R. M. M., ed., *Plant Life in Aquatic and Amphibious Habitats*, Blackwell Scientific, Oxford, pp. 53–76.

Schnitzler, A., Carbiener, R., and Trémolières, M., 1992. Ecological segregation between closely related species in the flooded forests of the Upper Rhine plain. *New Phytologist* 121:292–301.

Schwegler, T. and Brändle, R., 1991. Ethylenabhängige Wachstums- und Entwicklungsprozesse bei Stecklingen der Brunnenkresse (*Nasturtium officinale*). *Botanica Helvetica* 101:135–140.

Shipley, B. and Keddy, P. A., 1988. The relationship between relative growth rate and sensitivity to nutrient stress in twenty-eight species of emergent macrophytes. *Journal of Ecology* 76:1101–1110.

Shipley, B., Keddy, P. A., and Lefkovitch, L. P., 1991. Mechanisms producing plant zonation along a water depth gradient: a comparison with the exposure gradient. *Canadian Journal of Botany* 69:1420–1424.

Smirnoff, N. and Crawford, R. M. M., 1983. Variation in the structure and response to flooding of root aerenchyma in some wetland plants. *Annals of Botany* 51:237–249.

Smits, A. J. M., 1994. Ecophysiological Studies on Nymphaeid Water Plants. Ph.D. Thesis, University of Nijmegen, 198 pp.

Sykora, K. V., Scheper, E., and Van der Zee, F., 1988. Inundation and the distribution of plant communities on Dutch river dikes. *Acta Botanica Neerlandica* 37:279–290.

Thompson, K. and Grime, J. P., 1979. Seasonal variation in the seed bank of herbaceous species in ten contrasting habitats. *Journal of Ecology* 67:893–921.

Van den Brink, F. W. B., Van der Velde, G., and Bij de Vaate, A., 1993a. Ecological aspects, explosive range extention and impact of a mass invader, *Corophium curvispinum* Sars, 1895 (Crustacea: Amphipoda), in the Lower Rhine (The Netherlands), *Oecologia* 93:224–232.

Van den Brink, F. W. B., De Leeuw, J. P. M., Van der Velde, G., and Verheggen, G. M., 1993b. Impact of hydrology on the chemistry and phytoplankton development in floodplain lakes along the Lower Rhine and Meuse. *Biogeochemistry* 19:103–128.

Van der Meijden, E., Klinkhamer, P. G. L., de Jong, T. J., and Van Wijk, C. A. M., 1992. Meta-population dynamics of biennial plants: how to exploit temporary habitats. *Acta Botanica Neerlandica* 41:249–270.

Van der Sman, A. J. M., 1992. Flooding Resistance and Life Histories of Short-Lived Floodplain Herbs. Ph.D. Thesis, University of Nijmegen, 111 pp.

Van der Sman, A. J. M., Voesenek, L. A. C. J., Blom, C. W. P. M., Harren, F. J. M., and Reuss, J., 1991. The role of ethylene in shoot elongation with respect to survival and seed output of flooded *Rumex maritimus* L. plants. *Functional Ecology* 5:304–313.

Van der Sman, A. J. M., Joosten, N. N., and Blom, C. W. P. M., 1993a. Flooding regimes and life-history characteristics of short-lived species in river forelands. *Journal of Ecology* 81:121–130.

Van der Sman, A. J. M., Blom, C. W. P. M., and Barendse, G. W. M., 1993b. Flooding resistance and shoot elongation in relation to developmental stage and environmental conditions in *Rumex maritimus* L. and *Rumex palustris* Sm. *New Phytologist* 125:73–84.

Van de Steeg, H. M., 1984. Effects of summer inundation on flora and vegetation of river foreland in the Rhine area. *Acta Botanica Neerlandica* 33:365–366.

Van der Toorn, J. and Pons, T. L., 1988. Establishment of *Plantago lanceolata* L. and *Plantago major* L. among grass. II. Shade tolerance of seedlings and selection on time of germination. *Oecologia* 76:341–347.

Van der Valk, A. G., 1981. Succession in wetlands: A Gleasonian approach. *Ecology* 62:688–696.

Van Donselaar, J., 1961. On the vegetation of former river beds in The Netherlands. *Wentia* 5:1–85.

Venable, D. L., 1984. Using intraspecific variation to study the ecological significance and evolution of plant life-histories, in Dirzo, R. and Sarukhan, J., eds., *Perspectives on Plant Population Ecology*, Sinauer Association, Sunderland, MA p. 166–187.

Voesenek, L. A. C. J., 1990. Adaptations of *Rumex* in Flooding Gradients. Ph.D. Thesis, University of Nijmegen, pp. 159.

Voesenek, L. A. C. J. and Blom, C. W. P. M., 1989. Growth responses of *Rumex* species in relation to submergence and ethylene. *Plant Cell and Environment* 12:433–439.

Voesenek, L. A. C. J., Harren, F. J. M., Bögemann, G. M., Blom, C. W. P. M., and Reuss, J., 1990. Ethylene production and petiole growth in *Rumex* plants induced by soil waterlogging. *Plant Physiology* 94:1071–1077.

Voesenek, L. A. C. J. and Blom, C. W. P. M., 1992. Germination and emergence of *Rumex* in river floodplains. I. Timing of germination and seedbank characteristics. *Acta Botanica Neerlandica* 41:319–329.

Voesenek, L. A. C. J. and Van der Veen, R., 1994. The role of phytohormones in plant stress: too much or too little. *Acta Botanica Neerlandica* 43:91–127.

Voesenek, L. A. C. J., Van Oorschot, F. G. M. M., Smits, A. J. M., and Blom, C. W. P. M., 1993a. The role of flooding resistance in the establishment of *Rumex* seedlings in river floodplains. *Functional Ecology* 7:105–114.

Voesenek, L. A. C. J., Banga, M., Thier, R. H., Mudde, C. M., Harren, F. J. M., Barendse, G. W. M., and Blom, C. W. P. M., 1993b. Submergence-induced ethylene synthesis, entrapment, and growth in two plant species with contrasting flooding resistances. *Plant Physiology* 103:783–791.

Walker, L. R., Zasada, J. C., and Chapin, F. S., III, 1986. The role of life history processes in primary succession on an Alaskan floodplain. *Ecology* 67:1243–1253.

Washitani, I. and Masuda, M., 1990. A comparative study of the germination characteristics of seeds from a moist tall grassland community. *Functional Ecology* 4:543–557.

Wilson, S. D., 1991. Plasticity, morphology and distribution in twelve lakeshore plants. *Oikos* 62:292–298.

Wilson, S. D. and Keddy, P. A., 1986a. Species competitive ability and position along a natural stress/disturbance gradient. *Ecology* 67:1236–1242.

Wilson, S. D. and Keddy, P. A., 1986b. Measuring diffuse competition along an environmental gradient: results from a shoreline plant community. *American Naturalist* 127:862–869.

Yamamoto, F. and Kozlowski, T. T., 1987a. Effects of flooding, tilting of stem and ethrel application on growth, stem anatomy, and ethylene production of *Acer platanoides* seedlings. *Scandinavian Journal of Forest Research* 2:141–156.

Yamamoto, F. and Kozlowski, T. T., 1987b. Effects of flooding of soil on growth, stem anatomy, and ethylene production of *Thuja orientalis* seedlings. *IAWA Bulletin* 8:21–29.

PRACTICAL CONSIDERATIONS FOR WETLAND IDENTIFICATION AND BOUNDARY DELINEATION

8

R. W. Tiner

ABSTRACT

Since the 1970s, the federal government in the United States has been increasingly more active in regulating construction in wetlands. During this time many states have similarly developed programs to control development in wetlands. These regulations have necessitated the establishment of standardized procedures to identify and delineate wetlands. These methods utilize one or more types of wetland indicators, including hydrophytic vegetation, hydric soils, and other indicators of periodic wetness associated with wetlands. Several methods have been used for wetland identification: (1) vegetation-based methods, (2) soil-based methods, (3) three-parameter methods (using plants, soils, and other signs of wetland hydrology), and (4) the primary indicators method (relying on unique features to indicate wetlands). This article reviews wetland indicators and how they have been used in these methods to identify and delineate wetlands. Wetland mapping is also discussed. Recommendations are offered on how to improve identification of wetlands for regulatory purposes.

Since the 1960s and 1970s, wetlands have received increased attention in the United States due to the passage of wetland laws by numerous states and enactment of the Federal Clean Water Act and amendments. These laws and their accompanying regulations have placed certain restrictions on the use of wetlands on both private property and public lands. For the first time, it became important to establish the boundaries or limits of wetlands on a piece of ground to determine the areal extent of government jurisdiction. In effect, these laws created a type of land use zoning program where permitted activities and exempted activities were allowed, and other uses were not. The purpose of these laws was either to protect wetlands from destructive projects that could be constructed on less environmentally

1-56670-147-3/96/$0.00+$.50
© 1996 by CRC Press, Inc.

harmful sites or to regulate certain uses of wetlands that would adversely affect the quality of the nation's waters. The former was largely the intent of state wetland laws, while the latter was a primary goal of the Federal Clean Water Act. Different approaches were developed to identify wetlands and their boundaries for these laws. The purpose of this chapter is to generally discuss these methods and to recommend some practical approaches for identifying and delineating wetlands.*

WHAT IS A WETLAND?

Wetlands encompass a wide array of "wet lands" called marshes, bogs, swamps, fens, pocosins, wet meadows, and other names. The diversity of wetlands is further evidenced by the number of definitions that have been developed for wetland inventories and for various laws and regulations. Wetland definitions for conducting wetland inventories are scientifically based, since these surveys aim to identify wet habitats. Legal definitions are grounded in scientific concepts, but may be broader or narrower depending on the interests to be protected or regulated.

From a legal standpoint, a wetland is whatever the law says it is. This has led some people to suggest that the definition of "wetland" is simply a policy question to be decided by politicians and administrators rather than by scientists (Kusler, 1992). Others argue that the definition of a wetland is a scientific question since, for example, identification of wetlands is largely based on analyzing vegetation, soils, and/or hydrology and requires training in biological and physical sciences; also, scientists and not politicians or administrators discovered functions of wetlands that are now highly valued by society. Once the universe of wetlands is defined by scientists, the role of politicians and administrators is to decide how best to regulate such areas to satisfy society's needs and interests. The foundation of all legal definitions of recent origin comes from scientific studies of marshes, swamps, and similar areas that have documented significant functions important to society. These studies have helped change society's view of wetlands from that of a wasteland to one of a valuable natural resource. Ecologists, botanists, biologists, and other concerned scientists undoubtedly assembled the necessary information to draft legal wetland definitions and, in most cases, actually wrote these definitions.

Despite differences in the actual wording of various wetland definitions, they have much in common (see Table 1 for examples). Most wetland definitions emphasize the presence and predominance of plants (hydrophytes) that grow in water or in periodically flooded or saturated soils. Some wetland definitions also include nonvegetated areas such as mudflats, rocky shores along the coast, and ponds (see U.S. Fish and Wildlife Service definition in Table 1), while others

---

* At the request of the U.S. federal government, the National Academy of Sciences established a Committee on Characterization of Wetlands in late 1993 to review the scientific basis for wetland delineation. They have reviewed existing wetland delineation manuals and procedures and presented their findings and recommendations in a report: *Wetlands: Characteristics and Boundaries* (1995).

Table 1   Definitions of "Wetland" According to Selected Federal Agencies
           and State Statutes

Organization (reference)	Wetland definition
U.S. Fish and Wildlife Service (Cowardin *et al.*, 1979)	"Wetlands are lands transitional between terrestrial and aquatic systems where the water table is usually at or near the surface or the land is covered by shallow water. For purposes of this classification wetlands must have one or more of the following three attributes: (1) at least periodically, the land supports predominantly hydrophytes; (2) the substrate is predominantly undrained hydric soil; and (3) the substrate is nonsoil and is saturated with water or covered by shallow water at some time during the growing season of each year."
U.S. Army Corps of Engineers (Federal Register, July 19, 1977) and U.S. Environmental Protection Agency (Federal Register, December 24, 1980)	Wetlands are "those areas that are inundated or saturated by surface or ground water at a frequency and duration sufficient to support, and that under normal circumstances do support, a prevalence of vegetation typically adapted for life in saturated soil conditions. Wetlands generally include swamps, marshes, bogs and similar areas."
U.S.D.A. Soil Conservation Service (National Food Security Act Manual, 1988	"Wetlands are defined as areas that have a predominance of hydric soils and that are inundated or saturated by surface or ground water at a frequency and duration sufficient to support, and under normal circumstances do support, a prevalence of hydrophytic vegetation typically adapted for life in saturated soil conditions, except lands in Alaska identified as having a high potential for agricultural development and a predominance of permafrost soils."
State of Rhode Island Coastal Resources Mgmt. Council (RI Coastal Resources Mgmt. Program as amended June 28, 1983)	"Coastal wetlands include salt marshes and freshwater or brackish wetlands contiguous to salt marshes. Areas of open water within coastal wetlands are considered a part of the wetland. Salt marshes are areas regularly inundated by salt water through either natural or artificial water courses and where one or more of the following species predominate:" (8 indicator plants listed). "Contiguous and associated freshwater or brackish marshes are those where one or more of the following species predominate:" (9 indicator plants listed).
State of Rhode Island Dept. of Environmental Mgmt. (RI General Law, Sections 2-1-18 *et seq.*)	Fresh water wetlands are defined to include, "but not be limited to marshes; swamps; bogs; ponds; river and stream flood plains and banks; areas subject to flooding or storm flowage; emergent and submergent plant communities in any body of fresh water including rivers and streams and that area of land within fifty feet (50') of the edge of any bog, marsh, swamp, or pond." Various wetland types are further defined on the basis of hydrology and indicator plants, including bog (15 types of indicator plants), marsh (21 types of plants), and swamp (24 types of indicator plants plus marsh plants).
State of New Jersey (NJ Pinelands Comprehensive Management Plan, 1980; Pinelands Protection Act - NJ STAT. ANN. Section 13:18-1 to 13:29.)	"Wetlands are those lands which are inundated or saturated by water at a magnitude, duration and frequency sufficient to support the growth of hydrophytes. Wetlands include lands with poorly drained or very poorly drained soils as designated by the

Table 1   Definitions of "Wetland" According to Selected Federal Agencies and State Statutes (continued)

Organization (reference)	Wetland definition
	National Cooperative Soils Survey of the Soil Conservation Service of the United States Department of Agriculture. Wetlands include coastal wetlands and inland wetlands, including submerged lands." "Coastal wetlands are banks, low-lying marshes, meadows, flats, and other lowlands subject to tidal inundation which support or are capable of supporting one or more of the following plants:" (29 plants are listed.) "Inland wetlands" are defined as including, but not limited to, Atlantic white cedar swamps (15 plants listed), hardwood swamps (19 plants specified), pitch pine lowlands (10 plants listed), bogs (12 plants identified), inland marshes (6 groups of plants listed), lakes and ponds, and rivers and streams.
State of New Jersey (Coastal Wetland Protection Act-NJ STAT ANN. Section 13:9A-1 to 13:9A-10)	"Coastal wetlands" are "any bank, marsh, swamp, meadow, flat or other low land subject to tidal action in the Delaware Bay and Delaware River, Raritan Bay, Sandy Hook Bay, Shrewsbury River including Navesink River, Shark River, and the coastal inland waterways extending southerly from Manasquan Inlet to Cape May Harbor, or at any inlet, estuary or those areas now or formerly connected to tidal whose surface is at or below an elevation of 1 foot above local extreme high water, and upon which may grow or is capable of growing some, but not necessarily all, of the following:" (19 plants are listed.) Coastal wetlands exclude "any land or real property subject to the jurisdiction of the Hackensack Meadowlands Development Commission...."
State of Connecticut (CT General Statutes, Sections 22a–36 to 45, inclusive, 1972, 1987)	"Wetlands mean land, including submerged land, which consists of any of the soil types designated as poorly drained, very poorly drained, alluvial, and floodplain by the National Cooperative Soils Survey, as may be amended from time to time, of the Soil Conservation Service of the United States Department of Agriculture. Watercourses are defined as rivers, streams, brooks, waterways, lakes, ponds, marshes, swamps, bogs, and all other bodies of water, natural or artificial, public or private."
State of Connecticut (CT General Statutes, Sections 22a–28 to 35, inclusive 1969)	"Wetlands are those areas which border on or lie beneath tidal waters, such as, but not limited to banks, bogs, salt marshes, swamps, meadows, flats or other low lands subject to tidal action, including those areas now or formerly connected to tidal waters, and whose surface is at or below an elevation of one foot above local extreme high water."

include deepwater habitats, such as lakes, as wetlands (see New Jersey Pinelands Protection Act definition in Table 1). Deepwater habitats may have been included because a state wanted to regulate alternative uses of water bodies and saw the wetland bill as a convenient opportunity to safeguard these important habitats. After all, many wetlands are inextricably linked to a permanent water body.

Despite the inclusion of deepwater habitats in some definitions for vegetated wetlands, there is much commonality in the definitions.

INDICATORS OF VEGETATED WETLANDS

Many wetlands are dominated by plant species that grow only in wetlands. These species, called "obligate hydrophytes", are the best vegetative indicators of wetlands. These wetlands are readily identified by their flora. The average citizen can usually recognize these types of wetlands without much training. Many wetlands, however, lack the presence of these species and cannot be simply identified as wetlands by vegetation alone. They can be recognized by soil properties characteristic of wetlands. Thus, plants and/or soils are typically the most useful indicators of wetlands. These features are usually most applicable in situations where drainage has not been improved to effectively drain former wetlands. In these altered conditions, an evaluation of the effect of drainage must be performed to establish the absence or presence of wetland for regulatory purposes.

Hydrophytic Vegetation Indicators

Hydrophytes are plants that grow, or are capable of growing, in water or on a substrate that is periodically anaerobic (oxygen deficient) due to excessive water content (Tiner, 1991a). It is the presence of these plants that has been traditionally used to identify wetlands. Such plants have adapted physiologically and/or morphologically, or in other ways, to survive in and successfully colonize these water-stressed environments. Table 2 lists some mechanisms by which plants have successfully acclimated to these conditions. For reviews of these mechanisms, see Blom and others (1990), Crawford (1983), Gill (1970), Hook (1984), Hook and Scholtens (1978), Hook and others (1988), Jackson and Drew (1984), Kozlowski (1984), Teskey and Hinckley (1978), and Whitlow and Harris (1979). A response of a plant to flooding may be quite different than its response to waterlogging. Hosner (1958) found red ash (*Fraxinus pennsylvanica*) to be more tolerant of flooding than eastern cottonwood (*Populus deltoides*), but he subsequently found the latter species to be more tolerant of soil saturation (Hosner, 1958). This clearly demonstrates that caution must be exercised when extrapolating results of flood tolerance studies to conclude that one species is more water tolerant that another. Moreover, this is further complicated by the likely occurrence within a given species of distinct populations with genotypic or phenotypic differences in flood tolerance, as reported by Gill (1970), Keeley (1979), and Crawford and Tyler (1969).

Wetland plant communities have been called "hydrophytic vegetation" for identifying regulated wetlands (Environmental Laboratory, 1987; Sipple, 1988; Federal Interagency Committee for Wetland Delineation, 1989). A review of wetland definitions in Table 1 finds that some definitions include a list of plants that are examples of hydrophytes, whose presence should indicate wetland. If the

Table 2   Plant Adaptations or Responses to Flooding and Waterlogging

Morphological adaptations/responses	Other adaptations/responses
Stem hypertrophy (e.g., buttressed tree trunks)	Seed germination under water
Large air-filled cavities in center (stele) of roots and stems	Viviparous seeds
Aerenchyma tissue in roots and other plant parts	Root regeneration (e.g., adventitious roots)
Hollow stems	Growth dormancy (during flooding)
Shallow root systems	Elongation of stem or petioles
Adventitious roots	Root elongation
Pneumatophores (e.g., cypress knees)	Additional cell wall structures in epidermis or cortex
Swollen, loosely packed root nodules	Root mycorrhizae near upper soil surface
Lignification and suberization (thickening) of root	Expansion of coleoptiles (in grasses)
Soil water roots	Change in direction of root or stem growth (horizontal or upward)
Succulent roots	Long-lived seeds
Aerial root-tips	Breaking of dormancy of stem buds (may produce multiple stems or trunks)
Hypertrophied (enlarged) lenticels	
Relatively pervious cambium (in woody species)	
Heterophylly (e.g., submerged vs. emergent leaves on same plant)	
Succulent leaves	

Physiological adaptations/responses
Transport of oxygen to roots from lenticels
 and/or leaves (as often evidenced by oxidized
 rhizospheres)
Anaerobic respiration
Increased ethylene production
Reduction of nitrate to nitrous oxide and
 nitrogen gas
Malate production and accumulation
Reoxidation of NADH
Metabolic adaptations

From Tiner, R. W., *BioScience*, 41:236–247, 1991. With permission.

plant lists were presented in this table, anyone familiar with plant ecology would recognize some species that are exclusive to wetlands, such as smooth cordgrass (*Spartina alterniflora*) and skunk cabbage (*Symplocarpus foetidus*), and others that also occur in terrestrial habitats (uplands), such as red maple (*Acer rubrum*) and eastern hemlock (*Tsuga canadensis*). The former plants are the best vegetative indicators of wetlands due to their strict dependence on wetlands, whereas the latter are not, by themselves, useful indicators without considering associated species or other factors. Since many wetland plant communities are comprised of plant species that also grow in uplands, it is difficult to simply determine the presence of certain wetlands by vegetation alone.

Plant ecologists have long realized that many plant species have either broad ecological tolerances that allow them to successfully establish colonies in a variety of habitats or that ecotypes have evolved that are better adapted for colonizing varying habitats (e.g., wet, dry, strongly saline, fresh, sandy, or calcareous) (Tiner, 1991a). Turesson (1922a, 1922b, 1925) aptly demonstrated the existence of ecotypes within a given species. Ecotypes are populations or groups of populations having distinct genetically based morphological and/or physiological traits. Ecotypes of a given species are usually prevented from interbreeding by ecological barriers (Barbour *et al.*, 1980). The majority of plants occurring in

Table 3 Wetland Indicator Categories of Plant Species Under Natural Conditions

Wetland indicator category	Estimated probability of occurrence in wetlands	Estimated probability of occurrence in nonwetlands
Obligate wetland (OBL)	>99%	<1%
Facultative wetland (FACW)	67–99%	1–33%
Facultative (FAC)	34–66%	34–66%
Facultative upland (FACU)	1–33%	67–99%
Upland (UPL)	<1%	>99%

wetlands have either broad ecological amplitudes or have adaptive ecotypes (see Tiner, 1991a for detailed discussion of the concept of a hydrophyte).

The first and only comprehensive lists of hydrophytes for the U.S. were compiled by the U.S. Fish and Wildlife Service with support and cooperation from three other federal agencies (Army Corps of Engineers, Environmental Protection Agency, and the Soil Conservation Service). National, regional and state lists are now available (e.g., Reed, 1988; Tiner *et al.*, 1995). These lists reference plant species that have been found in U.S. wetlands.

Given that the affinity for wetlands varies considerably among plant species, the species on these lists have been separated into four "wetland indicator categories" that reflect differences in the expected frequency of occurrence in wetlands: (1) obligate wetland (OBL), (2) facultative wetland (FACW), (3) facultative (FAC), and (4) facultative upland (FACU) (see Table 3 for definitions). The national list contains 6,728 species out of a total of approximately 22,500 vascular plant species that exist in the U.S. and its territories and possessions (Reed, 1988). Only 31% of the nation's flora occur in wetlands often enough to be recorded on the list. Thus, the majority of the nation's plant life is virtually intolerant of prolonged flooding and saturation associated with wetlands. Of those species occurring in wetlands, only 27% are OBL species (Tiner, 1991a). The majority of the listed species, therefore, grow both in wetlands and nonwetlands to varying degrees. This fact clearly complicates wetland determinations and delineations based solely on analysis of vegetation, and was probably the main reason for developing a three-parameter method for wetland identification which requires examining vegetation, soils, and hydrology. The latter approach considers plant communities dominated by OBL, FACW, and/or FAC species as positive indicators for hydrophytic vegetation. With the inclusion of FAC species, many upland (terrestrial) plant communities have a positive vegetation indicator for wetland.

The OBL and FACW species are reliable indicators of wetlands in their natural undrained condition, with the OBL species being the best indicators. OBL species are usually characteristic of the wetter (seasonally flooded to permanently flooded) wetlands, but some species, such as Nuttall Oak (*Quercus nuttallii*), may occur only at the drier end of the moisture gradient. Examples of OBL species are presented in Table 4. Some plant families or genera are exclusive to wetlands, while most have certain species that are wetland dependent. The predominance of these species or their occurrence at some level of moderate abundance should reveal the presence of wetland. Some common wetland types that can be identified

Table 4   Examples of Obligate Hydrophytes That Are Widespread or Particularly Common In Certain Wetland Types In the United States. Genera Listed Contain All or Mostly Obligates

Aquatics

Alternanthera philoxeroides (Alligatorweed), *Azolla* spp. (Mosquito-ferns), *Brasenia schreberi* (Water-Shield), *Cymodocea filiformis* (Manatee-grass), *Eichornia crassipes* (Water Hyacinth), *Elodea* spp. (Water-weeds), *Hydrocotyle ranunculoides* (Water Pennywort), *Isoetes* spp. (Quillworts), *Lemna* spp. (Duckweeds), *Limnobium spongia* (American Frog-bit), *Myriophyllum* spp. (Water-milfoils), *Najas* spp. (Naiads), *Nuphar* spp. (Pond Lilies), *Nymphaea* spp. (Water Lilies), *Proserpinaca* spp. (Mermaid-weeds), *Ruppia maritima* (Widgeon-grass), *Thalassia testudinum* (Turtle-grass), *Utricularia* spp. (Bladderworts), *Vallisneria americana* (Wild Celery), *Zannichellia palustris* (Horned Pondweed), *Zostera marina* (Eel-grass).

Emergents (Herbs)

Alisma spp. (Water-plantains), *Calla palustris* (Wild Calla), *Caltha palustris* (Marsh Marigold), *Carex aquatilis* (Water Sedge), *Carex lasiocarpa* (Woolly Sedge), *Carex stricta* (Tussock Sedge), *Cicuta maculata* (Water Hemlock), *Decodon verticillatus* (Water-willow), *Drosera* spp. (Sundews), *Dulichium arundinaceum* (Three-way Sedge), *Eleocharis* spp. (Spike-rushes), *Eriophorum* spp. (Cotton-grasses), *Glyceria* spp. (Manna Grasses), *Hibiscus moscheutos* (Rose Mallow), *Iris versicolor* (Blue Flag), *Juncus canadensis* (Canada Rush), *Juncus roemerianus* (Black Needlerush), *Kosteletzkya virginica* (Seashore Mallow), *Leersia oryzoides* (Rice Cutgrass), *Lindernia dubia* (Water Pimpernel), *Lythrum lineare* (Salt Marsh Loosestrife), *Osmunda regalis* (Royal Fern), *Peltandra virginica* (Arrow Arum), *Polygonum hydropiperoides* (Water Pepper), *Polygonum sagittatum* (Arrow-leaved Tearthumb), *Pontederia cordata* (Pickerelweed), *Sabatia stellaris* (Annual Salt Marsh Pink), *Sagittaria* spp. (Arrowheads), *Salicornia virginica* (Perennial Glasswort), *Scirpus americanus* (Olney's Three-square), *Scirpus atrovirens* (Green Bulrush), *Scirpus validus* (Soft-stemmed Bulrush), *Sium suave* (Water Parsnip), *Solidago patula* (Rough-leaved Goldenrod), *Solidago uliginosa* (Bog Goldenrod), *Spartina alterniflora* (Smooth Cordgrass), *Spartina cynosuroides* (Big Cordgrass), *Symplocarpus foetidus* (Skunk Cabbage), *Triglochin* spp. (Arrow-grasses), *Typha* spp. (Cattails), *Woodwardia virginica* (Virginia Chain Fern), *Xyris* spp. (Yellow-eyed Grasses), *Zizania aquatica* (Wild Rice), *Zizaniopsis miliacea* (Giant Cutgrass).

Shrubs

Andromeda glaucophylla (Bog Rosemary), *Batis maritima* (Saltwort), *Betula pumila* (Bog Birch), *Borrichia frutescens* (Sea Ox-eye), *Cephalanthus occidentalis* (Buttonbush), *Chamadaphne calyculata* (Leatherleaf), *Forestiera acuminata* (Swamp Privet), *Kalmia polifolia* (Bog Laurel), *Lonicera oblongifolia* (Swamp Fly-honeysuckle), *Myrica gale* (Sweet Gale), *Rosa palustris* (Swamp Rose), *Salix caroliniana* (Swamp Willow), *Salix sericea* (Silky Willow), *Vaccinium macrocarpon* (Big Cranberry).

Trees

Avicennia germinans (Black Mangrove), *Carya aquatica* (Water Hickory), *Chamaecyparis thyoides* (Atlantic White Cedar), *Fraxinus caroliniana* (Carolina Ash), *Fraxinus profunda* (Pumpkin Ash), *Gleditsia aquatica* (Water Locust), *Nyssa aquatica* (Water Gum), *Planera aquatica* (Planer-tree), *Quercus lyrata* (Overcup Oak), *Rhizophora mangle* (Red Mangrove), *Salix nigra* (Black Willow), *Taxodium distichum* (Bald Cypress).

Vines

Smilax walteri (Red-berried Greenbrier).

simply by vegetation include tidal and nontidal marshes, mangrove swamps, Atlantic white cedar swamps, pocosins, fens, sedge meadows, bogs, cypress-gum swamps, and red maple-skunk cabbage swamps.

Plant communities lacking OBL species may be dominated by hydrophytes, but these species are more wide-ranging in their habitats and the hydrophytic (wetland) populations need to be determined by proving that they are growing in water or on a substrate that is at least periodically anaerobic due to excess wetness.

For these communities, the underlying soils will usually provide this supportive evidence.

Hydric Soil Indicators

Certain soil properties typically develop under reducing soil conditions associated with prolonged inundation and/or soil saturation (Vepraskas, 1992). The presence of organic soils and gleyed soils (hydric soils) have been recently used for wetland identification and delineation (Tiner and Veneman, 1987; Environmental Laboratory, 1987; Federal Interagency Committee for Wetland Delineation, 1989; Vepraskas, 1992; U.S.D.A. Soil Conservation Service, 1994).

Organic soils (Histosols, except Folists) have formed under conditions of nearly permanent flooding and/or soil saturation. The presence of thick deposits of peat or muck, therefore, is an excellent indicator of wetlands. Such deposits are generally indicative of the wetter wetlands. The vegetation growing on such soils is often dominated by obligate hydrophytes and many of these wetlands are easily recognized by their characteristic vegetation. On other organic soil sites, OBL species, while not dominant, are usually common enough to also clearly identify these wetlands by their vegetation.

Where surface water is present for extended periods during the growing season, a shallow deposit of organic material may form on the surface of wetlands. When this layer is 8 to 16 in. thick on top of a mineral soil, it is called a "histic epipedon". It is diagnostic of a wetland and may be useful for identifying wetland plant communities dominated by FACW species or where OBL species are present but not particularly abundant. A thin layer (e.g., greater than 1 in. thick) of muck or peat on top of sandy soils is also a useful wetland indicator in many areas.

Many wetland soils in temperate regions, however, lack a mucky or peaty surface layer. These mineral soils, however, typically possess dominant low-chroma (gleyed) colors in the subsurface layer (subsoil). The presence of a gleyed subsoil (B-horizon or C-horizon) immediately below the surface layer (A-horizon) typically indicates wetland. This property is useful for identifying many drier-end wetland communities lacking OBL species. Gleyed soils are formed by reduction which mobilizes iron, thereby causing iron to be translocated out of the soil and moved to another layer, or further downslope where it may precipitate as iron oxide mottles (redox concentrations) or as iron oxides in stream water. This loss of iron and manganese is commonly called redox depletion and results in the grayish (low chroma) colors typical of hydric mineral soils (Vepraskas, 1992). Iron oxides (e.g., reddish brown, yellowish, or orange in color) may form in gleyed soils. These "high chroma mottles" or redox concentrations typically indicate a fluctuating water table and significant oxidation in the affected layer.

Gleyed soils are characteristically grayish or dull in color (see Tiner and Veneman, 1987; Tiner, 1988; Tiner, 1991c; or Vepraskas, 1992 for color photographs of gleyed soils). These low chroma colors typically indicate significant reduction in the affected soil layer (horizon). Gleyed soils may also be grayish,

Table 5   Recommended List of Primary Indicators of Wetlands in the United States. The Presence of Any of These Characteristics in an Area That Has Not Been Significantly Drained Typically Indicates Wetland. The Upper Limit of Wetland is Determined by the Point at Which None of These Indicators are Observed

Vegetation Indicators of Wetland

V1. OBL species comprise more than 50% of the abundant species of the plant community. (*An abundant species is a plant species with 20% or more areal cover in the plant community.*)

V2. OBL *and* FACW species comprise more than 50% of the abundant species of the plant community.

V3. OBL perennial species collectively represent at least 10% areal cover in the plant community and are evenly distributed throughout the community and not restricted to depressional microsites.

V4. One abundant plant species in the community has one or more of the following morphological adaptations: pneumatophores (knees), prop roots, hypertrophied lenticels, buttressed stems or trunks, and floating leaves. (*Note:* Some of these features may be of limited value in the tropics.)

V5. Surface encrustations of algae, usually blue-green algae, are materially present. (*Note:* This is a particularly useful indicator of drier wetlands in arid and semiarid regions.)

V6. The presence of significant patches of peat mosses (*Sphagnum* spp.) along the Gulf and Atlantic Coastal Plain. (*Note:* This may be useful elsewhere in the temperate zone.)

V7. The presence of a dominant groundcover of peat mosses (*Sphagnum* spp.) in boreal and subarctic regions. (Indicator species will need to be designated by regional experts.)

Soil Indicators of Wetland

S1. Organic soils (except Folists) present.

S2. Histic epipedon (e.g., organic surface layer 8–16 in. thick) present.

S3. Sulfidic material (H_2S, odor of "rotten eggs") present within 12 in. of the soil surface.

S4. Gleyed[a] (low chroma) horizon or dominant ped faces (chroma 2 or less with mottles *or* chroma 1 or less with or without mottles) present immediately (within 1 in.) below the surface layer (A- or E-horizon) *and* within 18 in. of the soil surface.

S5. Nonsandy soils with a low chroma matrix (chroma of 2 or less) within 18 in. of the soil surface *and* one of the following present above the low chroma matrix and within 12 in. of the surface:
a. Iron and manganese concretions or nodules; or
b. Distinct or prominent oxidized rhizospheres along several living roots; or
c. Low chroma mottles.

S6. Sandy soils with one of the following present:
a. Thin surface layer (1 in. or greater) of peat or muck where a leaf litter surface mat is present; or
b. Surface layer of peat or muck of any thickness where a leaf litter surface mat is absent; or
c. A surface layer (A-horizon) having a low chroma matrix (chroma 1 or less and value of 3 or less) greater than 4 in. thick; or
d. Vertical organic streaking or blotchiness within 12 in. of the surface; or
e. Easily recognized (distinct or prominent) high chroma mottles occupy at least 2% of the low chroma subsoil matrix within 12 in. of the surface; or
f. Organic concretions within 12 in. of the surface; or
g. Easily recognized (distinct or prominent) oxidized rhizospheres along living roots within 12 in. of the surface; or
h. A cemented layer (orstein) within 18 in. of the soil surface.

S7. Native prairie soils with a low chroma matrix (chroma of 2 or less) within 18 in. of the soil surface *and* one of the following present:
a. Thin surface layer (at least $1/4$ inch thick) of peat or muck; or
b. Accumulation of iron (high chroma mottles, especially oxidized rhizospheres) within 12 in. of the surface; or
c. Iron and manganese concretions within the surface layer (A-horizon, mollic epipedon); or
d. Low chroma (gray-colored) matrix or mottles present immediately below the surface layer (A-horizon, mollic epipedon) and the crushed color is chroma 2 or less.

Soil Indicators of Wetland (continued)

S8. Remains of aquatic invertebrates are present within 12 in. of the soil surface in nontidal pothole-like depressions.

S9. Other regionally applicable, field-verifiable soil properties resulting from prolonged seasonal high water tables.

Note: Exceptions may occur as they do with any method and will be specified in the future as detected. Primary indicators for hydric prairie soils are based on field-tested recommendations by Dr. J. L. Richardson, North Dakota State University.

* Gleyed colors are low chroma colors (chroma of 2 or less in aggregated soils and chroma 1 or less in soils not aggregated; plus hues bluer than 10Y) formed by excessive soil wetness; other nongleyed low chroma soils may occur due to (1) dark-colored materials (e.g., granite and phyllites), (2) human introduction of organic materials (e.g., manure) to improve soil fertility, and (3) podzolization (natural soil leaching process in acid woodlands where a light-colored, often grayish, E-horizon or eluvial-horizon develops below the A-horizon; these uniform light gray colors are not due to wetness).

Modified from Tiner, R. W., *Wetlands,* 13:50–64, 1993. With permission.

because iron is present in its reduced form (ferrous iron, Fe^{2+}). Reduced soils may change color slightly within 30 minutes or less, upon exposure to air when ferrous iron is present (Vepraskas, 1992). They may appear more bluish at first and then change to a more dull blue-gray color. A colorimetric test using a-a diphyridal can also be used to confirm the presence of ferrous iron. Other soil properties may also be useful indicators of wetlands (see Table 5).

The U.S.D.A. Soil Conservation Service (SCS) has prepared a list of hydric soils for the country. Hydric soils are flooded, ponded, or saturated at a frequency and duration sufficient to create anaerobic conditions in the upper part of the soil (U.S.D.A. Soil Conservation Service, 1991). The intent of the definition of hydric soil was to identify soils that supported the growth of hydrophytes (Mausbach, 1994). The list of hydric soils may be useful for interpreting information in published soil survey reports, but the field indicators of hydric soil indicators are more significant and essential for wetland delineation. A list of these indicators has been published (U.S.D.A Soil Conservation Service, 1994a) and is currently being field tested. These indicators, once validated, should be extremely useful for determining the presence of wetlands in their natural, undrained conditions.

Wetland Hydrology Indicators

Certain methods for identifying wetlands require using indicators, other than vegetation and soil, to document the presence of wetland. The presence of surface water and interstitial soil water within the major portion of the plant's root zone during the growing season has been used as an indicator of wetland hydrology following federal wetland delineation methods (Environmental Laboratory, 1987; Sipple, 1988; Federal Interagency Committee for Wetland Delineation, 1989). Indirect indicators of the presence of water have also been used (Table 6). While these indicators provide evidence that an area is presently wet or that flooding or soil saturation has occurred, most fail to indicate the frequency or the duration of that event or the timing of the event. Only hydric soil properties, obligate hydrophytes, and plant morphological adaptations provide such evidence and their use

Table 6 List of Wetland Hydrology Indicators Used in U.S.
Federal Wetland Delineation Manuals (CE - Corps of
Engineers Manual, EPA - Environmental Protection
Agency Manual, FICWD - Federal Interagency
Committee for Wetland Delineation Manual)

Hydrology indicator	Manuals
Inundation during growing season	CE, EPA, FICWD
Soil saturation within 12″ of the surface during growing season	CE, EPA, FICWD
Water marks	CE, EPA, FICWD
Drift lines	CE, EPA, FICWD
Water-borne sediment deposits	CE, EPA, FICWD
Drainage patterns within wetlands	CE, FICWD
Hydrology data from published soil survey reports (after verifying soil type)	EPA
Surface scoured areas (includes bare areas subject to prolonged inundation)	EPA, FICWD
Moss lines on trees and shrubs	EPA
Morphological plant adaptations to inundation/soil saturation[a]	EPA, FICWD
Oxidized rhizospheres along living roots and rhizomes[b]	FICWD
Water-stained leaves	FICWD
Hydric soil characteristics (in areas with no apparent significant hydrologic modification - drainage)	FICWD

[a] CE manual considers these as indicators of hydrophytic vegetation.
[b] EPA manual considers these as indicators of hydric soil.

is limited to hydrologically unaltered sites. Despite these limitations, all of these hydrologic indicators have been used to verify that an area or plant community is presently subjected to wetland hydrological conditions following federal wetland delineation procedures.

METHODS TO IDENTIFY AND DELINEATE WETLANDS

A host of techniques have been developed to identify and delineate wetlands subject to government regulations. Certain agencies have conducted mapping projects to locate these areas, while others have developed methods to be employed on-the-ground to identify regulated wetlands. These approaches are briefly described in the following subsections.

Field Delineation Techniques

Four basic methods have been used for identifying and delineating wetlands in the field: (1) vegetation-based techniques, (2) soil-based methods, (3) three-parameter methods, and (4) the primary indicators method. Each of these approaches is briefly discussed below; for more detailed reviews, see Tiner (1989, 1993a, 1993b).

Vegetation-Based Methods

Vegetation-based methods require analysis of the plant community to identify regulated wetlands. These methods were the earliest techniques used to determine the jurisdictional limits of state wetland regulations. These methods were probably developed due to the involvement of botanists and wetland ecologists in creating wetland laws and the existence of much published technical and nontechnical botanical information, including field guides for plant identification on wetlands.

A "50% rule" was commonly employed to establish the predominance of wetland plants. Wetlands were identified where more than 50% of the plants were wetland species. This approach probably worked well for identifying salt marshes and the wetter freshwater nontidal wetlands where OBL and FACW species predominate, but it was less useful for identifying drier-end wetlands and wetland boundaries in areas of low relief.

Major shortcomings of this method include the lack of a comprehensive list of wetland plants and standardized methods for assessing the vegetation. The latter could have been easily resolved, since there are numerous techniques available for performing quantitative assessments of vegetation patterns in the published literature on plant ecology. The former shortcoming has been recently overcome by the development of national, regional, and state lists of plant species that occur in wetlands by the U.S. Fish and Wildlife Service (FWS). Despite these developments, there remain serious obstacles to using plants alone to identify wetlands: (1) the species level of plant taxonomy is not adequate for identifying wetland ecotypes (hydrophytes) of species that occupy both wetland and upland habitats, (2) many species growing in wetlands have broad ecological tolerances and are also associated with and even dominant in drylands, and (3) many wetland communities lack OBL species. As a result, any attempt to rely solely on vegetation for identifying wetlands will either fail to recognize all wetlands (error by omission) or will include nonwetlands as wetland (error by commission).

Soil-Based Methods

Soil-based methods rely on the presence of certain soil properties (e.g., hydric soils) to designate wetlands. These methods have not been widely used, probably due to little input from soil scientists in the wetland protection and regulatory process and a general lack of published material describing the utility of soils for identifying and delineating wetlands. In Connecticut and New Hampshire, soil scientists have contributed significantly to the development of state and local wetland protection programs and, as a result, soils are used to identify wetlands at the state level in Connecticut and in many municipalities with local wetland zoning ordinances in New Hampshire. Connecticut considers all poorly drained, very poorly drained, alluvial, and floodplain soils as wetlands (Table 1). This approach includes areas of nonhydric soils on floodplains as "wetland", which makes it more expansive than the conventional concept of wetland.

Some limitations of relying solely on hydric soil properties to designate wetlands include: (1) current techniques require considerable technical expertise in soil taxonomy in order to use; (2) the lack of "field guides" to aid and standardize hydric soil/wetland determinations presently leads to varied interpretations, especially in drier-end wetlands and at the wetland border in low-gradient systems; (3) the need to separate effectively drained hydric soils from other hydric soils since soil morphology usually does not significantly change; and (4) recognition that many types of hydric soils lack distinguishing morphologic properties for separating them from nonhydric soils without considering other factors (e.g., vegetation). The development of "Field Indicators of Hydric Soils of the United States" (U.S.D.A. Soil Conservation Service, 1994a) is the federal government's first attempt at standardizing hydric soil determinations through the use of field indicators. After field testing and verification, these verification indicators coupled with regional field guides illustrating these properties (e.g., Tiner and Veneman, 1987) could greatly facilitate the use of soils for wetland determinations by non-soil scientists as well as providing consistent interpretations by soil scientists.

Three-Parameter Methods

Three-parameter methods typically require verifying the presence of hydrophytic vegetation, hydric soils, and wetland hydrology to identify and delineate wetlands. The federal government developed this approach to identify wetlands subject to regulation under the Federal Clean Water Act. Three manuals using this type of approach have been developed; one by the Corps of Engineers (CE) (Environmental Laboratory, 1987), another by the Environmental Protection Agency (EPA) (Sipple, 1988), and a third by an interagency committee representing CE, EPA, FWS, and SCS (Federal Interagency Committee for Wetland Delineation, 1989). The first two manuals were prepared solely for identifying federally regulated wetlands, while the latter manual was developed for a broader purpose, including but not limited to wetlands potentially subject to federal regulation.

The 1989 interagency manual is intended to develop a standard scientifically based method to identify all vegetated wetlands in the U.S., regardless of their values or current regulatory programs and government policies. It was perceived that such a document would have broad utility and could be adopted by states and local governments interested in wetland protection and regulation independent of that implemented by the federal government. Subsequently, several states (e.g., New Hampshire, Maine, and Pennsylvania) adopted this manual for their state regulatory programs.

While all three manuals require making observations of vegetation, soils, and hydrology, the CE manual essentially requires finding positive indicators of all three parameters (hydrophytic vegetation, hydric soils, and wetland hydrology) for wetland identification and delineation (Environmental Laboratory, 1987). Tiner (1993b) provides an overview of this manual and mentions some of its limitations.

The EPA manual (Sipple, 1988) and the 1989 manual (Federal Interagency Committee for Wetland Delineation, 1989) both require consideration of all three parameters, but in many cases would accept less than three indicators as necessary for a wetland determination. For example, if a plant community satisfied both the hydrophytic vegetation and hydric soil criteria and no field indicators of wetland hydrology were present, the 1989 manual considered this area as wetland *provided it was not hydrologically altered* (e.g., drained). In this case, the vegetation and soils were deemed sufficient to make a wetland determination. Three-parameter fundamentalists criticized this protocol.

Some shortcomings of the three-parameter approaches include: (1) time required to perform wetland delineations, (2) the need to find positive indicators of wetland hydrology for hydrologically unmodified sites dominated by hydrophytic vegetation growing on hydric soils (CE manual), (3) applying one wetland hydrology standard to all wetlands, (4) the questionable strength and significance of some of the wetland hydrology indicators (e.g., drift lines, water marks, and "wetland" drainage patterns), (5) the use of nonstandard terms and concepts (e.g., drainage classes) in the hydric soil criterion, (6) too much reliance on professional judgment and room for individual interpretation or bias (CE manual), and (7) use of FAC species as indicators of hydrophytic vegetation.

Primary Indicators Method

The primary indicators method (PRIMET) is an outgrowth of traditional methods of identifying wetlands (Tiner, 1993a). It attempts to use vegetation patterns, soil properties, and other features that are unique to wetlands as diagnostic for wetland identification and delineation. The basic premise is that in the absence of significant hydrologic modification, these unique features can be reliably used to make wetland determinations. It further recognizes that significantly hydrologically altered sites require an assessment of the current hydrology because the preexisting soil and vegetation characteristics are persistent in most cases, and are no longer useful indicators of wetlands in such disturbed sites. Florida and Rhode Island have adopted this type of method for identifying regulated wetlands (Matthews, 1994).

Wetlands and their boundaries are defined by the presence of any one of numerous primary indicators (see Table 5 for examples). This approach is a rapid assessment technique which permits wetland determinations to be made with minimal investment of time. In doing so, it does not require detailed documentation of plant communities or soil characteristics. Also, proper use of the method requires that an initial evaluation of potential hydrologic modification be performed. Protocols for doing such evaluation need to be developed. Standardized procedures for handling hydrologically altered sites need to be developed, but this is true for all methods in current use. The PRIMET does, however, recognize that wetland hydrology requirements differ among wetland types and need to be considered when evaluating significantly drained sites.

Wetland Mapping

Wetland maps provide information on the location, type, and distribution of wetlands in a format that is available to and readily understood by the general public. It requires an enormous effort on behalf of the government, but it has the distinct advantage of showing people where these areas are located. Interested people can, therefore, determine the presence or absence of wetlands on their properties and get a good idea of the general location of these potentially regulated areas. Similar mapping has been done throughout the U.S. when private land is zoned for specific purposes (e.g., town zoning maps).

It must be readily acknowledged that wetland mapping is not as accurate as field delineation of wetland boundaries. Yet, conventional wetland mapping techniques (i.e., photointerpretation), combined with extensive field work to verify the maps, may be capable of producing a product that identifies the spatial extent of wetlands to the degree necessary to preserve the wetland ecosystem functions that the regulating agencies are interested in protecting. Aerial photographs of some wetlands (e.g., certain evergreen forested wetlands) are difficult to interpret (Tiner, 1990) and these wetlands may have to be addressed by other means, whereas most marshes, swamps, fens, and bogs are readily photointerpreted. Field verification of wetland boundaries for assessing site-specific project impacts will still, however, be necessary to establish the line on the ground where projects are encroaching on wetlands.

Wetland maps are produced through remote sensing techniques in two main ways: (1) photointerpretation of aerial photos, and (2) satellite image processing. Although satellite technology is improving, it is still not capable of producing as accurate and detailed wetland inventories as prepared through conventional photointerpretation techniques. The Federal Geographic Data Committee (1992) recently reached this conclusion in their report "Application of Satellite Data for Mapping and Monitoring Wetlands — Fact Finding Report". Most large-scale wetland inventories have used or are utilizing aerial photointerpretation techniques to produce wetland maps.

Several states have produced maps of regulated wetlands. The states of Connecticut, New Jersey, New York, Delaware, and Maryland have produced such maps for coastal wetlands (salt and brackish water marshes, and other tidal wetlands). These wetlands are among the most easily recognized through photointerpretation. New York has produced statewide inland wetland maps that show designated regulated wetlands.

Other states and the federal government have conducted wetland inventories and produced wetland maps. These maps do not identify the limits of government jurisdiction, but do show areas where permits may be required. For the most part, these maps are conservative in the identification of wetlands, with limited field work performed.

The most readily available wetland maps for the United States are National Wetlands Inventory (NWI) maps produced by the U.S. Fish and Wildlife Service (Figure 1). The NWI mapping techniques involve: (1) stereoscopic photointerpretation

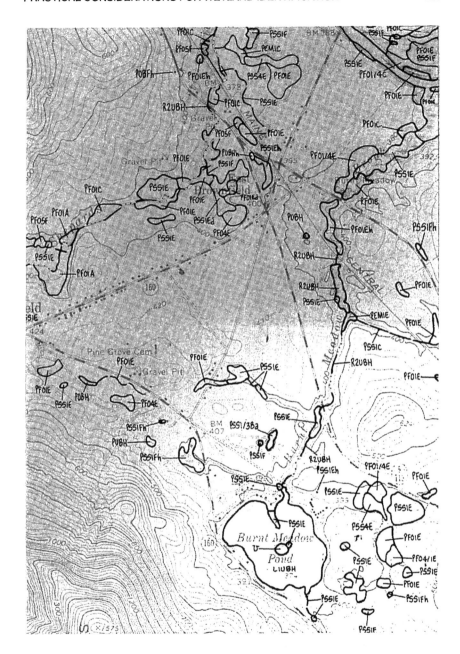

Figure 1   Example of a portion of the National Wetlands Inventory map for Brownfield, Maine (scale 1:24,000). Alpha-numeric codes represent different wetland and deepwater habitats, e.g., forested wetlands (PFO1E, PFO1/4C, PFO4E, PFO1A, etc.), scrub-shrub wetlands (PSS1E, PSS1F, PSS4E, PSS1/3Ba), emergent wetlands (PEM1E), ponds (PUBHh, PUBFh), lakes (L1UBH), and rivers (R2UBH). Minimum mapping unit is 1–3 acres, since 1:58,000 aerial photos were interpreted for this area.

of high- to medium-altitude aerial photography, (2) selective ground truthing, (3) review of existing information, (4) conventional cartographic procedures to produce a series of 1:24,000 maps, and (5) digital map database construction. U.S. Geological Survey topographic maps serve as the base maps for displaying wetlands inventory data. Minimum mapping units (mmu) of designated wetlands vary depending largely on the scale of the photographs used for wetland interpretation: 3–5 acres (1:80,000), 1–3 acres (1:58,000), and 1 acre (1:40,000). Certain conspicuous wetlands (e.g., prairie pothole marshes) and ponds smaller than the mmu may be shown. The wetlands which are more difficult to interpret from air photos (e.g., evergreen forested wetlands and seasonally saturated meadows and swamps) may be missed or conservatively mapped. In most areas, minimal field verification has been performed, so wetland mapping is conservative. More recently, however, use of 1:40,000 color infrared photography combined with extensive field verification have greatly improved the comprehensiveness and accuracy of the mapping.

The State of Maryland is on the cutting edge of wetland mapping technology. The Water Resources Administration is producing digital wetland maps at 1:7200 scale on orthophoto base images (Burgess, 1993). The maps do not show the boundaries of state-regulated wetlands, since Maryland requires site-specific wetland delineation for proposed projects. Instead, the maps are used as regulatory guidance maps showing the general limits of wetlands potentially subject to regulation. These maps are perhaps the most detailed and spatially accurate of any wetland maps produced to date for a large geographic area. Their mapping technique involves several steps: (1) stereoscopic photointerpretation of 1:40,000 color infrared aerial photography following NWI conventions, (2) extensive field verification, (3) vectorization of photointerpreted data (conversion to digital file), and (4) creation of digital orthophoto quarter/quad wetland maps. Figure 2 shows an example of a portion of one of these maps. The costs of this effort for Maryland is estimated at $4.5 million, which seems too expensive for most states and the nation as a whole. Yet, much of this cost is for producing base maps (orthophoto quarter-quads) and not for the compilation of wetland data. The base maps serve many purposes besides the wetland mapping, so the real cost of the wetland mapping is considerably less. Cooperative federal-state projects to produce orthophoto quarter-quads will further reduce costs for participating agencies. The U.S. Geological Survey is actively seeking cooperators for such projects.

DISCUSSION

Wetland maps have played and continue to play an important role in wetland protection. Many states have produced maps showing the location of regulated coastal wetlands, but few states have ventured to do this for inland wetlands. Wetland mapping has many advantages over field delineation from the standpoint of the regulated community. Most importantly, such maps could show the extent of government jurisdiction in a medium easily understood by most people. Presently,

Figure 2 Example of a state regulatory guidance wetland map (produced on a digital orthophoto quarter quad base) for Maryland. Alphanumeric codes for wetlands and deepwater habitats follow the U.S. Fish and Wildlife Service's wetland classification system.

the public in many areas must hire an environmental consultant to determine the limits of government jurisdiction (i.e., wetlands) on their property.

Despite these rather obvious advantages, why have not all regulatory agencies produced wetland maps? Wetland mapping can be expensive. To produce a set of regulatory maps similar to NWI maps, but with specially flown aerial photography, extensive field verification, and improved accuracy could cost an estimated $500 million to cover the entire U.S. (Don Woodard, U.S. Fish and Wildlife Service, personal communication). Yet, if one really considers the cost of the existing federal regulatory program, the cost of this mapping effort is not unthinkable in terms of the federal budget, especially when the cost is spread out over a number of years. The greatest concerns with wetland mapping probably are the limitations of remote sensing techniques for detecting and mapping all wetlands and that such techniques cannot delineate wetlands as well as a trained specialist can on the ground. Certain evergreen forested wetlands and drier-end wetlands (seasonally saturated and/or temporarily flooded) are difficult to identify and accurately delineate through remote sensing. Extensive field verification and consultation of existing data such as soil survey reports may, however, help overcome most of these technical problems. Regardless, any mapping effort will miss some wetlands, since by convention there is a minimum size limit that can effectively be shown on a map of a certain scale. Technical constraints and minimum mapping units will invariably result in the omission of some wetlands. Some key questions are (1) What types of wetlands are being missed? (2) What percent of the total wetland resource do they represent? (3) Are these wetlands vital to preserving the wetland functions and values that society desires and is interested in protecting? (4) If so, does this eliminate or greatly diminish the value of regulatory maps? and (5) Can maps be used to show the boundaries of, at least, certain wetland types that are readily identified through remote sensing techniques? The answers to these and other questions will largely determine the utility of maps for wetland regulation.

Given current remote sensing technologies and other available information (e.g., soil surveys), it is possible to produce a set of regulatory maps showing the location of water bodies, wetlands that are amenable to air photo interpretation, and well-defined nonwetlands (uplands), with the remaining lands designated as areas requiring field inspection to identify wetlands. The latter areas may be identified by considering landscape positions that favor wetland establishment and by consulting existing soil survey data. By separating "land" into three categories (wetland, upland, and land requiring field inspection), the geographic scope of potentially regulated land would be defined on a set of maps. Thus, the public would be duly informed of jurisdictional limits. Individuals looking for the most readily developable lands could simply consult the maps for optional parcels. This could result in a significant improvement in the efficiency of current wetland regulatory programs, in part by helping guide development away from wetlands to more suitable sites.

Mapping does not preclude the need for on-site inspections. Even where regulatory maps are produced, field delineation is still required to establish a line on the ground to guide landowners on where permits are necessary for work and where they are not, especially when projects are planned for construction in the wetland or near its border.

Standardization of field methods is needed to ensure accurate identification of wetlands and their boundaries. Such methods should be (1) technically sound by making use of current scientific knowledge to accurately identify wetlands, as well as being legally defensible (rather than being arbitrary and capricious); (2) precise enough to produce repeatable results so that different investigators would identify essentially the same boundary for a given wetland regardless of the time of year of field inspection; (3) practical and easy to use, emphasizing relatively easily observed features that can be recognized by generalists in major biological and physical sciences and not require highly specialized technical expertise to implement; (4) efficient — requiring only minimal effort to identify the wetter wetlands and increased effort for more difficult-to-identify wetlands; (5) capable of producing most determinations in a single site inspection; (6) able to permit wetland identification throughout the year (except perhaps when the soil is frozen and the area is snow covered); (7) sufficient in scope to encompass regional variation in wetlands throughout the United States; and (8) flexible enough to allow for limited use of professional judgment in difficult or confounding situations (Tiner, 1993a). Without standard methods and well-trained personnel to employ them, wetland identification and delineation would be extremely varied among individuals engaged in such tasks. This would pose a consistency problem for regulators and the regulated community alike. Moreover, it would further jeopardize protection of wetlands and their functions by failing to include them in the regulatory review process. Development of standardized wetland delineation methods and providing training to potential users are vital to the success of any wetland regulatory program.

In creating standard procedures for wetland delineation, the limitations of our knowledge of wetlands quickly become evident when considering wetland hydrology. The wetland paradox is that despite a wealth of information about wetlands, we do not know how wet a wetland is at its upper limit, or in other words, the minimum wetness required to create wetlands. There are no long-term studies of water table fluctuations along the soil moisture gradient between wetlands and uplands. Only recently have short-term studies been initiated (e.g., Allen *et al.*, 1989; Anderson *et al.*, 1980; Carter *et al.*, 1994; Roman *et al.*, 1985; Veneman and Tiner, 1990). This should not be construed as suggesting that we know nothing about wetland hydrology. Most, if not all, wetland ecologists would agree that an area flooded for a month or more during the growing season of each year is wet enough to support hydrophytes and be classified as wetland. Yet, is one week of flooding every other year sufficient for wetland establishment? Must wetlands be saturated to the surface for long periods and if so, how long, how often, and during what season? Is prolonged wetness during the "growing season" the most significant process affecting plant communities and wetland functions? How long does it take for soils to develop hydric properties? There are many other unanswered questions about wetland hydrology. Requiring verification of wetland hydrology for natural, undisturbed wetlands is unnecessarily burdensome and puts too much emphasis on a condition that is not well documented in the scientific literature (Environmental Defense Fund and World Wildlife Fund, 1992; Tiner, 1991b). Existing wetland definitions reflect this and do not mention specific time periods for inundation or soil saturation. Consequently, wetland identification has traditionally centered on plants and soils. These features are still the most useful indicators of wetlands in areas not significantly drained (Carter *et al.*, 1994; Federal Interagency Committee for Wetland Delineation, 1989; Sipple, 1985; Tiner, 1993a). As long as on-the-ground delineations are required, field indicators of wetland will be used for wetland identification and boundary delineation at sites with unaltered hydrology. While additional investigations are needed to corroborate use of certain indicators, there is little practical value to requiring a specific hydrology for natural wetlands in terms of days of flooding and/or soil saturation, given the absence of site-specific hydrologic data at most sites and a general lack of knowledge about the variations in hydrology between different wetland types and among similar types throughout the country.

Why then has there been so much recent attention in the U.S. placed on defining wetlands in terms of days of inundation and/or soil saturation? The need for this information stems from government regulatory programs that place certain restrictions on the use of wetlands on private property and the fact that many areas have experienced significant hydrologic modification through drainage ditches, tile drains, ground water withdrawals, river diversions, or other actions. In these highly disturbed areas, plant communities and soil properties are less reliable indicators of wetland, since they generally reflect previous hydrology. This is especially true of soils, which typically represent the best expression of long-term hydrology. Plants are more responsive to changing hydrologic conditions. A change in vegetation may indeed indicate altered hydrology (drainage), but the degree of the modification is usually not easily determined by vegetative analysis. Where UPL species have become dominant, there should be widespread agreement that the wetland is now effectively drained. Yet, in most cases, this does not happen, but instead, FACU species that occur in natural wetlands such as black cherry (*Prunus serotina*), may be establishing themselves, while most of the preexisting plant community remains, being able to tolerate the more mesic conditions created by drainage. The increase in these types of species is not definitive in determining the extent or effectiveness of drainage.

In sites with significantly altered hydrology, the current hydrology needs to be determined. This can be accomplished in several ways ranging from conducting on-site ground water well studies and interpreting stream gauge data (for floodplain sites) to modeling studies for determining the scope and effect of ditches and tile drains. The U.S.D.A. Soil Conservation Service has prepared a handbook to aid in determining wetland hydrology (U.S.D.A. Soil Conservation Service, 1994b). This manual is undergoing peer review and field testing.

Consideration and evaluation of the frequency and duration of flooding and soil saturation should be restricted to sites whose hydrology appears to be significantly altered (e.g., extensive drainage). For these situations, a minimum threshold of wetland hydrology needs to be developed to aid regulators in determining areas wet enough to potentially regulate. This threshold should vary according to the wetland type (region, climate, physiography, topography, etc.), since the minimum wetness for a prairie pothole wetland in the Upper Midwest should be different than that of a bog or a tidal marsh due to differing hydrologies. Compiling the best available information from the literature, with review by leading wetland scientists in each region, should allow reasonable and practical minimum standards to be developed. Again, such standards should be applied only to significantly disturbed sites and not to more natural wetlands or wetlands with minor drainage (e.g., a single ditch through a large wetland). Soil and vegetation indicators are reliable for identifying the latter wetlands. Regional committees, including federal and state wetland experts, could be established to expand the list of primary indicators for the variety of wetlands occurring in each region. The Primary Indicators Method provides the most practical and expeditious approach to identifying these wetlands. With delineation performed quickly and efficiently, investigators can then put more effort towards functional analysis of wetlands relative to the proposed alterations.

REFERENCES

Allen, S. D., Golet, F. C., Davis, A. F., and Sokoloski, T. E., 1989. Soil-Vegetation Correlation in Transition Zones of Rhode Island Red Maple Swamps. *U.S. Fish and Wildlife Service Biological Report* 89(8).

Anderson, K. L., Lefor, M. W., and Kennard, W. C., 1980. Forested wetlands in eastern Connecticut: their transition zones and delineation. *Water Resources Bulletin* 16:248–255.

Barbour, M. G., Burk, J. H., and Pitts, W. K., 1980. *Terrestrial Plant Ecology*. Benjamin Cummings Publ., Menlo Park, CA.

Blom, C. W. P. M., Bogemann, G. M., Laan, P., van der Sman, A. J. M., van de Steeg, H. M., and Voesenek, L. A. C. J., 1990. Adaptations to flooding in plants from river areas. *Aquatic Botany* 38:29–47.

Burgess, B., 1993. A detailed analysis of Maryland's digital orthophoto quarter quad and wetlands mapping programs. Maryland Department of Natural Resources, Water Resources Administration, Annapolis, MD. May 10, 1993 unpublished mimeo.

Carter, V., Gammon, P. T., and Garrett, M. K., 1994. Ecotone dynamics and boundary determination in the Great Dismal Swamp. *Ecological Applications* 4:189–203.

Committee on Characteristics of Wetlands, 1995. *Wetlands: Characteristics and Boundaries*. National Research Council, National Academy Press, Washington, D.C.

Cowardin, L. M., Carter, V., Golet, F. C., and LaRoe, E. T., 1979. Classification of Wetlands and Deepwater Habitats of the United States. *U.S. Fish and Wildlife Service* FWS/OBS-79/31. Wahington, D.C.

Crawford, R. M. M., 1983. Root survival in flooded soils. In: A. J. P. Gore, ed. *Ecosystems of the World*. 4A. Mires: Swamp, Bog, Fen, and Moor — General Studies. Elsevier, Amsterdam. pp. 257–283.

Crawford, R. M. M. and Tyler, P. D., 1969. Organic acid metabolism in relation to flooding tolerance in roots. *Journal of Ecology* 57:235–244.

Environmental Defense Fund and World Wildlife Fund., 1992. How Wet Is a Wetland? The Impacts of the Proposed Revisions to the Federal Wetlands Delineation Manual. Washington, D.C.

Environmental Laboratory, 1987. *Corps of Engineers Wetland Delineation Manual*. U.S. Army Engineer Waterways Experiment Station, Vicksburg, MS, Tech. Rep. Y-87-1.

Federal Geographic Data Committee, 1992. Application of Satellite Data for Mapping and Monitoring Wetlands — Fact Finding Report. Wetlands Subcommittee, FGDC, Washington, D.C. Tech. Rep. 1.

Federal Interagency Committee for Wetland Delineation, 1989. *Federal Manual for Identifying and Delineating Jurisdictional Wetlands*. U.S. Army Corps of Engineers, U.S. Environmental Protection Agency, U.S. Fish and Wildlife Service, and U.S.D.A. Soil Conservation Service, Washington, D.C.

Gill, C. J., 1970. The flooding tolerance of woody species — a review. *Forest Abstracts* 31:671–688.

Hook, D. D., 1984. Adaptations to flooding with fresh water. In T.T. Kozlowski, ed. *Flooding and Plant Growth*. Academic Press, Orlando, FL, pp. 265–294.

Hook, D. D. and Scholtens, J. F., 1978. Adaptations and flood tolerance of tree species. In D. D. Hook and R. M. M. Crawford, eds. *Plant Life in Anaerobic Environments*. Ann Arbor Scientific Publ., Ann Arbor, MI, pp. 229–331.

Hook, D. D., McKee, W. H., Jr., Smith, H. K., Gregory, J., Burrell V. G., Jr., DeVoe, M. R., Sojka, R. E., Gilbert, S., Banks, R., Stolzy, L. H., Brooks, C., Matthews, T. D., and Shear, T. H., eds. 1988. *The Ecology and Management of Wetlands. Vol. 1. Ecology of Wetlands*. Timber Press, Portland, OR.

Hosner, J. F., 1958. The effects of complete inundation upon seedlings of six bottomland tree species. *Ecology* 39:371–373.

Hosner, J. F., 1959. Survival, root and shoot growth of six bottomland tree species following flooding. *Journal of Forestry* 59:927-928.

Jackson, M. B. and Drew, M. C., 1984. Effects of flooding on growth and metabolism of herbaceous plants. In: T. T. Kozlowski, ed. *Flooding and Plant Growth*. Academic Press, New York, pp. 47–128.

Keeley, J. E., 1979. Population differentiation along a flood frequency gradient: physiological adaptations to flooding in *Nyssa sylvatica*. *Ecological Monographs* 49:89–108.

Kozlowski, T. T., 1984. Responses of woody plants to flooding. In: T. T. Kozlowski, ed. *Flooding and Plant Growth*. Academic Press, New York, pp. 129–163.

Kusler, J. A., 1992. Wetlands delineation: an issue of science or politics? *Environment* 34:7–11.

Matthews, F. E., 1994. Ratification of wetland bill gives state a single, unified delineation method. *Florida Specifier* 16(5).

Mausbach, M. J., 1994. Classification of wetland soils for wetland identification. *Soil Survey Horizons* 35:17–25.

Reed, P. B., Jr., 1988. National List of Plant Species that Occur in Wetlands: National Summary. *U.S. Fish and Wildlife Service Biol. Rep.* 88(24), Washington, D.C.

Roman, C. T., Zampella, R. A., and Jaworski, A. Z., 1985. Wetland boundaries in the New Jersey pinelands: ecological relationships and delineation. *Water Resources Bulletin* 21:1005–1012.

Sipple, W. S., 1985. *Wetland Identification and Delineation Manual*. U.S. Environmental Protection Agency, Office of Federal Activities, Washington, D.C. Draft Report.

Sipple, W. S., 1988. Wetland Identification and Delineation Manual. Volume I. Rationale, *Wetland Parameters, and Overview of Jurisdictional Approach*. Volume II. Field Methodology. U.S. Environmental Protection Agency, Office of Wetlands Protection, Washington, D.C.

Teskey, R. O. and Hinckley, T. M., 1978. Impact of Water Level Change on Woody Riparian and Wetland Communities. Vol. V. Northern Forest Region, *U.S. Fish and Wildlife Service* FWS/OBS-78/88, Washington, D.C.

Tiner, R. W., 1993a. The primary indicators method — a practical approach to wetland recognition and delineation in the United States. *Wetlands* 13:50–64.

Tiner, R. W., 1993b. Field recognition and delineation of wetlands. Chapter 5. In: M. S. Dennison and J. F. Berry, eds. *Wetlands: Guide to Science, Law, and Technology*. Noyes Publications, Park Ridge, NJ, pp. 153–198.

Tiner, R. W., 1991a. The concept of a hydrophyte for wetland identification. *BioScience* 41:236–247.

Tiner, R. W., 1991b. How wet is a wetland? *Great Lakes Wetlands Newsletter* 2:1–4, 7.

Tiner, R. W., 1991c. Maine Wetlands and Their Boundaries. State of Maine, Department of Economic and Community Development, Office of Comprehensive Planning, Augusta, ME.

Tiner, R. W., Jr., 1990. Use of high-altitude aerial photography for inventorying wetlands in the United States. *Forest Ecology and Management* 33/34:593–604.

Tiner, R. W., Jr., 1989. Wetland boundary delineation. In: S. K. Majumdar, R. P. Brooks, F. J. Brenner, and R. W. Tiner, Jr., eds. *Wetland Ecology and Conservation: Emphasis in Pennsylvania*. Pennsylvania Academy of Sciences, c/o Lafayette College, Easton, PA, pp. 231–248.

Tiner, R. W., Jr., 1988. *Field Guide to Nontidal Wetland Identification*. Maryland Department of Natural Resources, Annapolis, MD, and the U.S. Fish and Wildlife Service, Newton Corner, MA.

Tiner, R. W. and Veneman, P. L. M., 1987. Hydric Soils of New England. University of Massachusetts Cooperative Extension Service, Amherst, MA, Bulletin C-183.

Tiner, R., Lichvar, R., Franzen, R., Rhodes, C., and Sipple, W., 1995. *1995 Supplement to the List of Plant Species that Occur in Wetlands: Northeast (Region 1)*. U.S. Fish and Wildlife Service, National Wetlands Inventory, St. Petersburg, FL. Supplement to Biol. Rpt. 88(26.1).

Turesson, G., 1922a. The species and the variety as ecological units. *Hereditas* 3:100–113.

Turesson, G., 1922b. The genotypical response of the plant species to the habitat. *Hereditas* 3:211–350.

Turesson, G., 1925. The plant species in relation to habitat and climate. *Hereditas* 6:147–236.

U.S.D.A. Soil Conservation Service, 1994a. *Field Indicators of Hydric Soils of the United States*. June, 1994. Washington, D.C.

U.S.D.A. Soil Conservation Service, 1994b. *Hydrology Tools for Wetland Determination*. Version 2.0. Washington, D.C.

U.S.D.A. Soil Conservation Service, 1991. *Hydric Soils of the United States*. Washington, D.C. Misc. Publication No. 1491.

U.S.D.A. Soil Conservation Service, 1988. *National Food Security Act Manual*. Washington, D.C.

Veneman, P. L. M. and Tiner, R. W., 1990. Soil-Vegetation Correlations in the Connecticut River Floodplain of Western Massachusetts. *U.S. Fish and Wildlife Service Biol. Rep.* 90(6), Washington, D.C.

Vepraskas, M. J., 1992. Redoximorphic Features for Identifying Aquic Conditions. North Carolina Agricultural Research Service, North Carolina State University, Raleigh. Tech. Bull. 301.

Whitlow, T. H. and Harris, R. W., 1979. Flood Tolerance In Plants: A State-of-the-art Review. U.S. Army Engineers Waterways Experiment Station, Vicksburg, MS. Tech. Rep. E-79-2.

9 TOWARD THE INTEGRATION OF WETLAND FUNCTIONAL BOUNDARIES INTO SUBURBAN LANDSCAPES

W. G. Pearsell and G. Mulamoottil

ABSTRACT

Municipal planners and engineers require a framework for assessing the potential impacts of suburban development proposals on wetlands that parallel the community and drainage planning process. The more common approach is to conduct an environmental impact assessment (EIA) close to the design stage of suburban developments, well after community and drainage planning have been completed. The suburban landscape is changed frequently by decisions made early in the planning process that may affect the spatial requirements of wetland functions and values beyond the physical boundary of the wetland. The *ecological* or *functional* boundaries may straddle *administrative* boundaries and further complicate the environmental assessment process.

This chapter examines the status of wetland planning in Edmonton, Alberta and describes problems with the present land use planning process and offers recommendations for the integration of administrative and ecological boundaries that can be applied broadly in developing suburban landscapes within a hierarchical planning system. It is emphasized that spatial requirements of wetland functions must be addressed early in the planning process.

INTRODUCTION

The City of Edmonton, Alberta has in recent years initiated procedures to formally recognize values of natural areas in suburban landscapes. The city has developed a number of criteria for classification of these areas into *natural areas*, *environmentally significant areas*, or *environmentally sensitive areas*. A recent study commissioned by the city (Geowest, 1994) has classified natural areas according to these criteria. Although there is no formal policy regarding the incorporation of natural areas into suburban landscapes, the City Council requested consideration of the values of these natural areas and the viability of incorporating them into new neighborhoods as a condition of approval. The Planning and Development Department has indicated that a formal policy will be forthcoming in the near future. Meanwhile, the city has issued a paper outlining its proposed policy direction for dealing with natural areas in the suburban land use planning process.

Presently, the City of Edmonton requires a system of hierarchical land use planning products for community and drainage planning of new suburban developments. The community planning products deal with the orderly establishment of neighborhoods. Neighborhoods function as administrative boundaries for a number of services such as parks, schools, shopping centers, and transportation. The drainage planning products are produced to ensure that stormwater management systems will function within neighborhoods as the landscape is changed from natural and/or agricultural uses to suburban land uses. Natural drainage boundaries are either incorporated into or altered to fit the drainage requirements of neighborhoods.

With the additional need to incorporate natural areas such as wetlands into neighborhoods, it is important to recognize ecological boundaries. Many wetland functions are influenced by the ecological gradient between the terrestrial and aquatic environments. Therefore, not only must the wetland boundary be recognized in development plans, but the establishment of boundaries for different wetland functions in order to incorporate wetlands into new neighborhoods is also required. The problem with much of wetland planning is that community and drainage planning decisions are made without considering the spatial requirements of wetlands beyond what is generally accepted as the physical boundary of a wetland. Any decisions that may negatively affect wetland functions and values run the risk of affecting the long-term viability and sustainability of wetlands within a suburban context.

This chapter focuses on the present land planning system in Edmonton and as an example describes land use planning (i.e., community, drainage, and wetland plans) that is required for suburban development approvals. Using Edmonton as a case study, some problems are identified related to the integration of administrative and ecological boundaries in the planning process. Recommendations are made on the type of products required to integrate the requirements of both ecological and administrative boundaries into a hierarchical land use planning system for suburban developments.

WETLAND PLANNING CONCEPTS

Functions and Values

Historically, the practice of wetland drainage has been deemed appropriate to increase the economic benefits of development projects. Wetlands are still often perceived as wasteland areas to be filled and eliminated to provide land suitable for development (Ontario Ministry of Natural Resources, 1989). However, these drainage practices and the reclamation of wetland areas do not consider their value to suburban development and the importance of the functions and values of wetlands within a drainage basin. The economic benefits of keeping wetlands in their natural state for sustainable uses such as hunting, fishing, and trapping practices are highlighted in a recent study by Elliot and Mulamoottil (1992).

Publications on wetland functions have shown that wetlands are important in regional hydrology and in providing functions such as ground water recharge and discharge, flood storage and flood desynchronization, shoreline anchoring and dissipation of erosive forces, sediment trapping, retention and removal of nutrients and pollutants, and habitat for fish and wildlife (de Witt and Solway, 1977; Good et al., 1978; Greeson et al., 1979; Federation of Ontario Naturalists, 1981; Adamus, 1983; Hook et al., 1988a, 1988b).

Wetland Boundary

The establishment of wetland boundaries is an important administrative step in protecting wetlands. Land use planning is required to ensure that development proposals are considered fairly. The lack of an effective method for wetland boundary delineation forces planners to evaluate development proposals that involve wetlands in an ad hoc manner. The United States has devoted considerable resources to develop an operational definition and evaluation method for determining wetland boundary (Federal Interagency Committee for Wetland Delineation, 1989).

The location of a wetland boundary depends upon the long-term hydrologic processes in a watershed (Gosselink et al., 1981, Pechmann et al., 1989; Weller, 1989; Koerselman et al., 1990). The hydrologic system to which a wetland is connected is sensitive to changes in the system such that changes in one part of the watershed or hydrologic setting may greatly affect another part of the system and is indeed a dynamic process (Carter et al., 1979; Phillips, 1989). Whether hydrologic changes are naturally occurring or anthropogenic, changes to the boundary will increase or decrease the size of a wetland, as the case may be.

Wetland Hydroperiod

The wetland hydroperiod is the pattern or "signature" of seasonal water levels and is unique to each wetland (Mitsch and Gosselink, 1986). The wetland hydroperiod is the "integration of all inflows and outflows of water and it is influenced by physical features of the terrain and by proximity to other bodies of water" (Mitsch and Gosselink, 1986). The hydroperiod is important in shaping the composition of vegetation and establishing the community structure. It is critical for many plant species (Zimmerman, 1988) and any changes to the hydroperiod have the potential to affect established vegetation patterns. Any changes to the vegetation have the potential to affect other wetland functions especially wildlife and wildlife habitat.

Wetland Functional Boundary

Gosselink *et al.* (1990) have asserted that because of the dynamic forces influencing wetland form there is a need to manage the impact of land use change on a landscape scale such that "the boundaries of an assessment unit encompass an area that is, to the extent possible, ecologically closed to the water and nutrient flows so that forces external to the basin can be minimized and large enough to satisfy the home range and habitat requirements of the farthest ranging animal species of interest."

Changes in a watershed can affect hydroperiods and concomitantly the seasonal requirements of wildlife. For example, Pechmann *et al.* (1989) examined the influence of the wetland hydroperiod on the diversity and abundance of metamorphosing juvenile amphibians for three wetlands in South Carolina. The life cycles of these amphibians starts with an aquatic larval stage that is followed by a metamorphosis and a final movement to terrestrial habitats as adults. Adults migrate annually from the terrestrial habitat to the wetlands to breed. Amphibian larvae must reach a minimum body size before they can undergo metamorphosis. If the wetland dries up before the adult stage in their life cycle, they may die or fall victim to predators. Therefore, a change in the wetland hydroperiod which increases the number of drying out periods directly influences the amphibian population.

A hydroperiod with more permanent water levels is not necessarily better because these kinds of wetlands can support more amphibian predators. The overall conclusion to be drawn from this discussion is that when an amphibian habitat is identified as an important wetland function, the maintenance of the wetland hydroperiod and provision of sufficient upland habitat beyond the wetland boundary are essential. This is, in effect, what must be regarded as the functional boundary. In order to protect the habitat functions, the functional boundary must be delineated and protected.

Changes to the hydroperiod caused by land use change may affect the overall functioning of the wetland over time. If this is the case, the location of the wetland boundary will also be affected. Any delineation of a wetland boundary requires the recognition of the long-term hydrologic processes so that the negative effects of development can be avoided or mitigated.

Figure 1 outlines a conceptual model of functional boundaries of a wetland. Four functional boundaries of a wetland (A, B, C, and D) that may be present in typical prairie marshes are represented. Functional Boundary A is outside of the drainage basin and is a recharge area for the wetland. Functional Boundaries B and C are boundaries for wildlife habitat requirements (e.g., duck nesting and amphibian habitat). Functional Boundary D represents the drainage boundary of the watershed and indicates the source area for water to the wetland. Any changes that affect the spatial characteristics of a functional boundary may negatively affect the functions of a particular wetland. It is important to note that there may be more than one functional boundary and that the functional boundaries do not necessarily coincide.

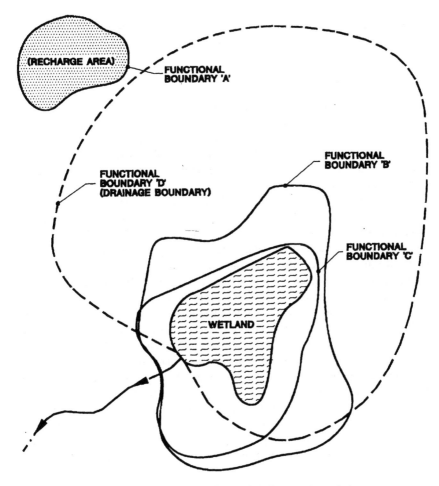

Figure 1   Conceptual model of wetland functional boundaries.

PLANNING FOR SUBURBAN DEVELOPMENT, EDMONTON ALBERTA

Community planning in Alberta is governed by the *Planning Act* which requires communities to prepare a number of different plans. In addition, the City of Edmonton has developed requirements for a number of other nonstatutory community, drainage, and to a lesser extent, environmental plans. The net effect is a sophisticated hierarchical system of planning tools at different spatial scales for community and drainage planning. However, wetland planning and integration into the hierarchical system of planning is not appropriately addressed.

The following section briefly outlines the various community planning, drainage planning, and wetland planning tools used by the City of Edmonton to plan suburban developments according to regional, district, and neighborhood scales. The following discussion does not consider the city's treatment of other natural areas such as woodlots. Table 1 is a summary of the discussion. The product of each process is identified in the table with the functional boundary of each plan.

Regional Scale Plans

Regional Planning Commissions are required to prepare a regional plan under Section 46 of the *Planning Act*. The *Edmonton Metropolitan Regional Plan*, prepared by the Edmonton Metropolitan Regional Planning Commission, encompasses an area greater than the City of Edmonton and provides for "the balanced, orderly and economic use of land in the Edmonton metropolitan region" for the communities in the plan area.

The *Edmonton General Municipal Plan (Bylaw 9076)* (1990) is a policy framework document for land use planning at the municipal level and provides a general framework for development. A General Municipal Plan (GMP) is required under Section 63 of the Alberta *Planning Act*. The purpose of Edmonton's GMP is "to identify strategic, city wide land use planning issues and to address these issues by providing objectives, policies and a framework for action." There are no specific policies regarding natural areas in the GMP.

The Regional Master Plan is a drainage concept plan that defines strategies and alternatives for storm and sanitary system extensions for the Edmonton area. This plan is prepared by the City of Edmonton Drainage Branch and is developed prior to Area Structure Plans. This plan is formulated on the watershed concept and alternatives for stormwater management (SWM) are defined. The Regional Master Plan focuses on management of stormwater for future developments.

A Watershed Plan is another drainage plan at a regional scale that is used to document existing constraints and best management alternatives for development within each watershed in the city. This plan is prepared by the City of Edmonton. The major emphasis is on posturbanization flow rates, quantities, and quality of water.

Wetland planning is not done at a district scale; however, the city has recently completed an inventory of natural areas in Edmonton (Geowest, 1994) and it is an important first step in developing a system for planning to integrate wetlands into a suburban landscape.

District Scale Plans

The main type of community plans at the district scale are Area Structure Plans (ASPs) and Neighborhood Area Structure Plans (NASPs). ASPs are statutory plans and provide a framework to plan for a number of neighborhoods. ASPs generally cover areas of at least 200 ha. Sometimes an NASP is prepared for one or two neighborhoods in instances where ASPs are not practicable. ASPs and NASPs identify where residential, commercial, institutional, and recreational sites are to be located. They also identify where municipal services such as water, sewer, roads, and fire protection services will be routed; and indicate the number of people expected to live in new areas and describe the staging. The GMP

Table 1 Summary of Land Use Planning in Edmonton: Products and Functional Boundaries at Regional, District, and Neighborhood Scales

Scale of planning	Community planning		Drainage planning		Wetland planning	
	Product	Functional boundary	Product	Functional boundary	Product	Functional boundary
Regional	General Municipal Plan	The City	Regional Master Plan	The City of Edmonton	Natural Area Inventory (existing)	Wetland and Wetland Complexes. Does not consider functional boundary
District	Area Structure Plan	A number of neighborhoods covering areas at least 200 ha	Watershed Plan or Preliminary Drainage Report	Watershed (can extend outside the city)	Addendum to Area Structure Plan	None specified
	Servicing Concept Design Briefs	Same as ASP but identifies significant natural features	Area Master Plan	Defined by Watershed Plan or the Preliminary Design Report		
Neighborhood	Neighborhood Structure Plan	An area that can support 4,000 to 7,000 people	Neighborhood Design Report	Defined by Area Master Plan	Natural Areas Assessment	None specified

policies apply in the district scale plans. In addition to ASPs and NASPs, the city administration may prepare a nonstatutory *Servicing Concept Design Brief* for large-scale, newly developing suburban communities.

Drainage Planning at the district scale is summarized in an Area Master Plan. The Area Master Plan is prepared to develop servicing schemes respecting the long-term user requirements and justifications for the selection of the solutions proposed, and defines the characteristics of selected alternatives for sanitary and storm drainage servicing. The Area Master Plan develops and proposes the optimum sewer and drainage servicing schemes which will meet short- and long-term servicing needs. Area Master Plans are prepared to provide technical input to Area Structure Plans.

If wetlands are present within an ASP, an addendum to the ASP is recommended by the planning and development department. This addendum should provide an assessment of the economic and ecological viability of the wetland in the context of the proposed development concept.

Neighborhood Scale Plans

The community planning at the neighborhood scale is summarized in a Neighborhood Structure Plan (NSP). A Neighborhood Structure Plan is a component or subunit of an Area Structure Plan. These plans are generally for areas that can support 4000–7000 people and have greater detail than ASPs. An NSP outlines the general pattern for subdivisions including land use type, transportation network, location and size of neighborhood facilities, and staging. It is adopted by Council as an amendment to the Area Structure Plan. NSPs are nonstatutory and must adhere to "spirit, intent and guidelines" of the ASPs. NSPs provide detail to provide the basis for detailed subdivision and zoning of the land. NSPs are prepared by the landowner/developer and are approved by Council as a plan bylaw.

The Neighborhood Designs Report (NDR) is prepared to define detailed design requirements for storm and sanitary sewer facilities required to service the development area. This report is prepared by the private developer. The NDR requires a sediment control plan and addresses the needs and defines measures to be implemented throughout the construction period and beyond to mitigate problems within the development and external to the development that affect downstream sewer systems and/or other developments.

When the Neighborhood Structure Plan is submitted, information concerning wetlands and other natural areas is required by the planning and development department. This information should include a reiteration of the information provided at the ASP level. If the ASP did not provide this information, the proponent will provide information of ESA/SNA sites and/or any information related to additional sites in the plan area. This assessment should consider the economic and ecological viability of a wetland in the proposed development concept.

The City of Edmonton Planning and Development Department is requesting that a separate Natural Areas Assessment be prepared that considers wetlands and other natural areas, whether or not they have been designated as special areas by the city.

INTEGRATING WETLAND FUNCTIONAL BOUNDARIES INTO SUBURBAN LANDSCAPES

The evolution of the suburban land use planning process in Edmonton to include wetlands and other natural areas is presenting new challenges to planners involved in the land development process. Suburban landscapes are often radically reshaped physically to accommodate associated changes in hydrology, and one of the biggest challenges is the incorporation of wetlands that are, after all, largely a manifestation of existing landscape processes such as climate, hydrology, nutrient fluxes, and topography. Land that is presently being developed in the Edmonton region was previously reclaimed for agriculture. There are very few unaltered landscapes in the City of Edmonton outside of the ravines and river valley. In other words, the *environmentally sensitive areas* and *environmentally significant areas* designated by the City of Edmonton are remnants of landscapes, but have been modified and influenced heavily by agricultural development.

Wetlands have also been subjected to alterations, with the loss of a number of wetland characteristics that are now valued by the City of Edmonton. It is reasonable to expect that wetlands will be further modified as a result of the type of activity associated with suburban development. Some of these changes may be positive and others will be negative.

Currently, wetlands and other natural areas and their boundaries are being evaluated after the administrative boundaries of Neighborhood Structure Plans and hydrological boundaries of the Neighborhood Design Report have been finalized. The decision to incorporate wetlands into suburban development has taken place at the end of a 20-year planning horizon. In some cases, decisions on the location of community infrastructure such as sewers, stormwater management facilities, road networks, public transit, school sites, and park sites have already been made.

It could be argued that the land use planning system will change over time as new Area Structure Plans are produced and wetland planning becomes an accepted component of the planning process. However, without recognizing the functions of wetlands and their values, as well as the spatial requirements to sustain these functions at the district planning scale (i.e., Area Structure Plan, Area Master Plan), decisions concerning infrastructures such as roads, sewers, and stormwater management lakes will have been made and investment decisions made accordingly. Detailed information on the requirements to sustain a wetland should be collected and used for district-scale planning decisions. It is argued that the functional boundary is a useful concept for organizing this information in a way that is compatible with community and drainage planning objectives.

Although administrative and functional boundaries do not necessarily coincide, a suburban planning system has been developed in Edmonton that integrates these boundaries through an iterative planning process, moving from a regional scale to a district scale to a neighborhood scale. To reach the implementation stage of this planning process, much time and effort must be expended by both the Planning and Development Department and the development industry to develop future neighborhoods that are acceptable to the needs of residents both in terms of the community and stormwater management functions. The challenge facing the City of Edmonton is the integration of functional boundaries into new suburban developments so that wetland functions and values are not compromised.

If wetlands are to be integrated into suburban neighborhoods, a parallel wetland planning system with products that correspond to the plans developed for both community and drainage plans at each of the regional, district, and neighborhood planning levels is required. Although the City of Edmonton is moving in this direction, it is to be expected that for the reasons outlined above the functional boundary will not be recognized early enough in the planning process. The following discussion outlines suggestions for wetland planning products at different scales.

Regional Scale Planning

As indicated earlier, a regional scale environmental inventory currently exists (Geowest, 1994). This inventory for the most part provides binary information on whether a wetland or other natural area is present and an initial evaluation of its significance. However, this inventory does not consider wetland functional boundaries and, therefore, there is no estimate of the area of land needed to sustain a wetland in a suburban setting.

This inventory should be part of the information used to prepare the future General Municipal Plans, Regional Master Plans, and Watershed Plans. In particular, wetland areas are significant constraints to be identified in the preparation of the Watershed Plan. In the past, low-lying areas have been identified as the best locations for stormwater management (SWM) lakes. The integration of wetlands into the suburban landscape will most likely preclude low-lying areas as sites for SWM lakes.

District Scale Planning

Wetland functions and values should be evaluated for all wetlands in a district which is designated for suburban planning at a district scale. This information is important to help in the decision making process used to develop the Area Master Plan and the Area Structure Plan. It is critical at this stage to establish wetland functional boundaries. The spatial extent of land that is required to support wetland functions and values, as for example waterfowl upland habitat and maintenance of wetland hydroperiod, should be assessed. It is also important at this stage to evaluate whether wetland functions and values are sustainable in a suburban landscape. For example, a waterfowl nesting habitat will not be compatible with family pets such as the dogs and cats that will be introduced into a subdivision.

Information related to functional boundaries is essential for the establishment of administrative boundaries for services such as parks, schools, shopping areas, arterial and collector roadways, and SWM lakes. It is important that wetland functional boundaries are established at this stage so that firm decisions about the overall form and structure of suburban neighborhoods are made before expectations about the ultimate developability of the land are established. Information concerning the wetland functional boundary should be part of the public hearing process required for Area Structure Plans.

Neighborhood Scale Planning

Wetland planning at the neighborhood scale, as proposed by the City of Edmonton, has limitations if spatial functions are to be maintained. It should be noted that this is the stage in the planning process that most areal decisions are made. The main wetland planning is completed after these decisions are made. Because of the land use changes that have been decided previously, such as changes to drainage boundaries, most wetland assessments will probably conclude that a wetland will not be viable in its current state in the context of suburban development because the hydrology has been changed. The reasons for this include:

- Stormwater management diversions;
- An increase in runoff volume and flows because of reduction in watershed permeability;
- Encroachment into the functional boundary by inappropriate land uses that destroy or negatively affect wildlife habitat;
- Alterations to the gradients (see Chapter 5, this volume).

The same limitations that apply to wetland planning at a neighborhood scale also apply to using Environmental Impact Assessments as a tool to consider ecological considerations for wetlands planning at the end of the decision making process.

CONCLUSIONS

It is imperative that planning decisions involving wetlands consider the spatial requirements or functional boundaries required to maintain an individual wetland function. Further, these requirements should be integrated into the planning process at the level of community and drainage plans at all scales of the planning process. Due consideration of wetland protection early in the planning process is required to prevent future losses of wetlands. However, how wetland protection is to be accommodated in the land use planning process is a vexing problem. Land use planning is an exercise in making informed choices and balancing quality of life issues with an ability to pay for the communities we live in. However, if questions about the preservation of wetland functions are not considered in the decision making process, informed choices cannot be made and viable future options will be lost.

REFERENCES

Adamus, R. R., 1983. A method for wetland functional assessment, volume 1, *Federal Highway Administrative Report* FHWA-Ip-82-23, Washington, D.C.

Bond, W. K., Cox, K. W., Heberlein, T., Manning, E. W., Witty, D. R., and Young, D. A., 1992. Wetland evaluation guide: final report of the wetlands are not wastelands project, North American Wetlands Conservation Council (Canada).

Carter, V., Bedinger, M. S., Novitzki, R. P., and Wilen, W.O., 1979. Water resources and wetlands, in P. E. Greeson, J. R. Clark, and J. E. Clark, eds. *Wetland Functions and Values: The State of our Understanding*, American Water Resources Association, Minneapolis, Minnesota, 344.

Cowardin, L. M., Carter, V., Golet, F. C., and LaRoe, E. T., 1979. Classification of wetlands and deepwater habitats of the United States, U.S. Fish and Wildlife Services Office of Biological Services, Washington, D.C., 1979.

de Witt, C. B. and Solway, E., eds., 1977. Wetlands: Ecology, Values, and Impacts, in Proceedings of the Waubesa Conference on Wetlands held in Madison, Wisconsin, June 2–5, 1977, Institute for Environmental Studies, University of Wisconsin-Madison, Madison, Wisconsin.

Elliot, L. and Mulamoottil, G., 1992. Agricultural and marshland uses on Walpole Island: profit comparisons. *Canadian Water Resources Journal* 17:111–119.

Federal Interagency Committee for Wetland Delineation, 1989. *Federal Manual for Identifying and Delineating Jurisdictional Wetlands,* U.S. Army Corps of Engineers, U.S. Environmental Protection Agency, U.S. Fish and Wildlife Services, and U.S.D.A. Soil Conservation Service, Washington, D.C.

Federation of Ontario Naturalists, 1981. In Anne Champagne, ed., *Proceedings of the Ontario Wetlands Conference*, hosted by the Federation of Ontario Naturalists and the Department of Applied Geography, Ryerson Polytechnical Institute. Federation of Ontario Naturalists, Don Mills, Ontario.

Geowest Environmental Consultants Ltd., 1994. Inventory of environmentally sensitive and significant natural areas, City of Edmonton.

Good, R. E., Whigham, D. F., and Simpson, R. L. eds., 1978. *Freshwater Wetlands*, Academic Press, New York.

Gosselink, J. G., Schaffer, G. P., Lee, L. C., Burdick, D. M., Childers, D. L., Leibowitz, N. C., Hamilton, S. C., Boumans, R., Cushman, D., Fields, S., Kock, M., and Visser, J. M., 1990. Landscape conservation in a forested wetland watershed: Can we manage cumulative impacts? *BioScience*, 40:588.

Gosselink, J. G., Bayley, S. E., Conner, W. H., and Turner, R. E., 1981. Ecological factors in the determination of riparian wetland boundaries, in J. R. Clark and J. Benforadok, eds., *Wetlands of Bottomland Hardwood Forests*, Elsevier Scientific, New York, p. 197.

Greeson, P. E., Clark, J. R., and Clark, J. E., eds., 1979. *Wetland Functions and Values: The State of Our Understanding*, American Water Resources Association, Minneapolis, Minnesota.

Hook, D. D., McKee, W. H., Jr., Smith, H. K., Gregory, J., Burrel V. G., Jr., Devoe, M. R., Sojka, R. E., Gilbert, S., and Shear, R., eds., 1988a. *The Ecology and Management of Wetlands, Volume 1: Ecology of Wetlands,* Timber Press, Portland, OR.

Hook, D. D., McKee, W. H., Jr., Smith, H. K., Gregory, J., Burrel V. G., Jr., Devoe, M. R., Sojka, R. E., Gilbert, S., and Shear, R., eds., 1988b. *The Ecology and Management of Wetlands. Volume 2: Management Use and Value of Wetlands,* Timber Press, Portland, OR.

Koerselman, W., Claessens, D., Ten Den, P., and Van Winden, E., 1990. Dynamic hydrochemical and vegetation gradients in fens, *Wetlands Ecology and Management*, 1:73.

Mitsch, W. J. and Gosselink, J. G., 1986. *Wetlands*, Van Nostrand Reinhold, New York.

Ontario Ministry of Natural Resources, 1989. Wetland Planning Policy Statement: Implementation Guidelines. Toronto.

Ontario Ministry of Natural Resources and Environment Canada, 1984. An Evaluation System for Wetlands of Ontario South of the Precambrian Shield, Second Edition, Wildlife Branch Outdoor Recreation Group, Ontario Ministry of Resources and Canadian Wildlife Service, Ontario Region Environmental Conservation Service, Environment Canada, Toronto.

Pechmann, J., Scott, D., Gibbons, J., and Semlitsch, R., 1989. Influence of wetland hydroperiod on diversity and abundance of metamorphosing juvenile amphibians, *Wetland Ecology and Management*, 1:3–11.

Phillips, J. D., 1989. An evaluation of the factors determining the effectiveness of riparian forest along a coastal plain river, *Journal of Hydrology*, 110:221.

Weller, M. W., 1989. Plant and water level dynamics in an east Texas shrub/hardwood bottomland wetland, *Wetlands*, 9:72.

Zimmerman, J. H., 1988. A multi-purpose wetland characterization procedure, featuring the hydroperiod, in Proceedings of the National Wetland Symposium: Wetland Hydrology, September 16–18, 1987, Chicago, Illinois, Association of State Wetland Managers, Inc., Berne, New York, 31.

10 MANAGEMENT GOALS AND FUNCTIONAL BOUNDARIES OF RIPARIAN FORESTED WETLANDS

J. E. Cox

ABSTRACT

A general definition of riparian zones as three-dimensional zones of direct interaction between terrestrial and aquatic ecosystems provides a starting point for consideration of the ecological structure and function of riparian forests needed for informed management. Some fundamental ecological issues include: (1) the scale of inquiry (landscape, site), (2) structure and function of the terrestrial system (corridors for plant and animal habitat and dispersal, biodiversity conservation, succession, and species composition), and (3) ecological functions for aquatic ecosystems. Rather than presenting criteria for delineating the boundaries of riparian forests, a case is presented for the importance of management goals in determining functional boundaries of riparian forested wetlands. When considering such forested wetlands with an ecosystem approach, there can not be one single boundary defined for riparian areas that is independent of human goals and independent of scale considerations.

INTRODUCTION

In southern Ontario there is a growing recognition that surface water and ground water supplies are under siege. One response by groups and agencies such as Trout Unlimited, school children, local environmental groups, municipalities, and conservation authorities is the initiation of stream bank rehabilitation and restoration projects at many different locations.

From an ecological perspective 'stream bank' areas are conceptually contained under the landscape concept of riparian zones: the "three-dimensional zones of direct interaction between terrestrial and aquatic ecosystems" (Gregory

1-56670-147-3/96/$0.00+$.50

et al., 1991). When the term "interaction" is based on a hydrological perspective, the proximity to streams with their low topographic position in the landscape suggests that riparian zones are usually wetlands. Also arising from their geographical context, riparian zones are ecotones. An ecotone generally is defined as a "zone of transition between adjacent ecological systems, having a set of characteristics uniquely defined by space and time scales and by the strength of the interactions between adjacent ecological systems" (Holland, 1988 as cited by Gosz, 1991). In this case the label ecotone is not intended to suggest that riparian areas have no unique characteristics. Their unique ecological characteristics as ecotones will be explored further below.

Riparian zones are not necessarily wetlands when 'interaction' is examined from a broader ecological perspective, for example, considering plant and animal habitat requirements and dispersal mechanisms. An example of this could be the cutbank at the outside edge of a stream meander. The trees on this bank may be elevated above the stream to the extent that they are not in a classifiable wetland, however, they are certainly providing inputs to the stream and pollen for trees on the opposite depositional bank which may be wetland. Given that there are other nonhydrological ecological connections, the term riparian *area* rather than zone is suggested as an alternative here.

A hydrology-based definition, where riparian area is synonymous with wetland, may be quite appropriate for relatively undisturbed areas. However, the landscape definition of riparian areas given above which can go beyond hydrology may be necessary for riparian corridors in urban settings (where they exist), which are frequently highly disturbed through development and subsequent use. Urban and near-urban areas are often the focus of management activities. There, the hydrology is often severely modified and community structures may more strongly reflect management and historical use patterns. Characteristics of those communities may productively be thought of as falling along a gradient of "more riparian" to "less riparian" as identified by the plants and animals found. The contrast between boundary determination at the site scale of communities, populations, and individuals vs. the landscape level will infuse the discussions that follow.

Boundaries and Management

While reviewing the ecology of riparian ecosystems, the issue of wetland boundaries is explored. Frequently, part of the decision making process for riparian areas involves explicit or implicit determination of functional boundaries. These boundaries may be those that are functional at the landscape scale, influencing large-scale hydrological patterns and relationships between forest patches. Smaller, site-scale boundaries may be functional in terms of recognizable changes in local plant communities in relation to biophysical site conditions and life history characteristics of local animal populations.

The need for boundary determination may be related to a number of issues including legislation of buffer widths, landowner decisions on buffer width, animal population management, species composition, and placement of restoration attempts and to the setting of clear goals and recognition of cost effective solutions for resource- and labor-demanding projects. In the context of southern Ontario, where forest was probably the dominant vegetation type along streams before European settlement, this chapter considers that while boundaries are a concept largely conceived in landscape terms, the delineation of riparian forest boundaries for management also involves smaller-scale natural community and population processes and human interactions and management goals.

ECOLOGY OF RIPARIAN FOREST ECOTONES

Large-Scale Processes: Ecology at the Landscape Scale

Important management issues arise from the characteristics of riparian areas as ecotones, as well as their position in the landscape. Both have resulted in riparian areas having an important function in the landscape, out of proportion to the actual area they occupy (Risser, 1990). For their relative size they are extremely important (Gregory et al., 1991).

Some theoretical studies have looked at riparian areas in terms of their functions as connecting corridors and buffers from surrounding development for maintaining regional biodiversity of animal and plant species (Naiman et al., 1993; Holland et al., 1990). As ecotones between streams and upland areas, riparian forests tend to have representative species typical both of swamp forest and upland forest, with the overall result that they contain more species than either the isolated forested wetland or upland forest. Complex microtopography tends to be characteristic of riparian areas and subsequently there is a wide range of microhabitats available. The numerous small-scale variations in topography and soils result from the lateral migration of the channel over time (Naiman et al., 1993). Furthermore, streams can play an important role in plant dispersal, moving plant propagules rapidly to colonize previously disturbed sites. As with herbivore movement, each species using a primarily riparian habitat will likely have a different functional boundary whose distance from the stream may vary through life cyle stages. For example, managing to encourage game animals may lead to very different boundaries than if amphibians are of primary interest.

Related to boundaries perpendicular to the stream, as opposed to the previous boundaries parallel to the stream, is the function of riparian areas as corridors. Riparian areas have the potential to link together otherwise isolated wetland and upland forest fragments in the landscape. This can be expected to facilitate the movement of species between suitable habitat areas. Therefore, boundaries in terms of corridor function are more closely linked to major vegetation units such as forest vs. field than to specific hydrological gradients. The connection of areas of habitat by corridors of similar functional composition can greatly enhance the survival of plants and animals over the levels which would be expected for isolated patches (Harker et al., 1993). Enhanced opportunities over small patches for finding and using resources and an expanded gene flow are among the important functions of corridors.

By the same argument, corridors can also facilitate the dispersal of predators and of pests and alien plant species. Therefore, decisions on altering the spatial extent and composition of riparian forests through ecological restoration should look at all potential consequences at both the local and landscape scale. The general importance of scale to biological systems has been recognized as having extremely important implications both for fundamental ecological understanding (Pickett *et al.*, 1989) and in the context of environmental planning and management. Dominant processes may differ as the scale of interest is larger or smaller.

Site-Scale Processes: Succession and Community Composition

Though not easy to clearly link to functional boundary determination, understanding successional processes for ecotones is essential for management. Many riparian plants are adapted to periodic flooding (Barnes, 1983; Blom *et al.*, 1990; Rosgen, 1993; Holland *et al.*, 1990) and waterlogged soils (Adamus *et al.*, 1990). In a systems ecological sense, feedback may exist between the plants adapted to flooding and flood water intensity. Forested riparian areas can substantially modify stream hydrology, decreasing the energy of flood waters, and maintaining low flows during dry periods (Armour *et al.*, 1991). In situations where flooding still occurs into riparian areas, both the extent of flooding and high water tables can be used to determine a meaningful boundary. Other abiotic soil and site conditions determine to a large extent which species will continue to survive at a given site over time.

For the American mid- and southwest and western Cordillera, considerable work has been done examining the relationships between riparian vegetation and site conditions (Tellman *et al.*, 1993; Sedell and Swanson, 1990; Baker, 1989; Reily and Johnson, 1981; Stromberg and Patten, 1990, to list a few). Those studies considering vegetation have focused on general characteristics of dispersal and response to environmental variation on more of a landscape and process scale. There is a glaring lack of such work for the temperate northeastern U.S. and central Canada. Exceptions are the unpublished work of Collette (1983) for herbaceous plants; the ecosystem studies at Hubbard Brook (Bormann and Likens, 1979) also included community characterization.

In addition to abiotic factors, biotic interactions such as competition may place major constraints on the development of vegetation in riparian areas. Herbivores, including beaver, can modify a considerable amount of vegetation in a short time (Barnes, 1983). The extent of herbivore movement through riparian areas and human land use patterns present real constraints to succession.

The role of chance events and different disturbances is increasingly recognized as important to the course of ecological succession in general (van der Maarel, 1988). That is, current theory does not suggest that for a given site there is only one, inevitable climax community. Rather, for any site, there may be a range of directions for change over time. This lack of determinism is magnified when not only "natural" disturbances are considered, such as strong wind and fire, but also human-initiated disturbances such as resource extraction, pollution inputs,

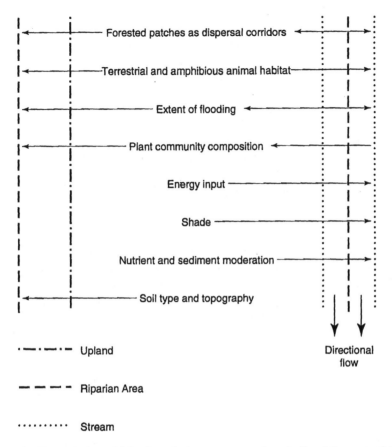

Figure 1   Conceptual model for important processes and controlling influences of the riparian area in the context of the landscape.

and heavy recreation. Such disturbances can greatly complicate boundary delineation, in that many of them will impact adjacent ecological areas regardless of functional boundaries. The dynamic nature of disturbance provides an argument for building into management systems either the flexibility to alter boundary classifications, or the allowance of a sufficiently large geographical area beside the streams to allow for inevitable changes in functional boundaries.

STRUCTURE AND FUNCTION FOR AQUATIC ECOSYSTEMS

The fundamental connection of streams to their vegetated banks has been explored from different theoretical perspectives. Connections which can influence functional boundaries range from the possible feedback between the vegetation form and the hydrology, discussed above, to energy and nutrient interactions (the river continuum concept of Vannote et al., (1980), and shading and buffer functions (Gregory et al., 1991) (Figure 1).

Riparian areas are often managed to buffer streams from excess nutrients, sediment (Correll, 1991; Dillaha *et al.,* 1988), and pollutants (NRC, 1992). Dissolved nutrients may be removed by uptake by riparian plants, or by soil microbial action, particularly in saturated soils (Peterjohn and Correll, 1984; Phillips, 1990; Hillbricht-Ilkowska and Pieczynska, 1993). Plant species vary in their effectiveness at nitrogen removal, with nitrogen-fixing plants being least effective (Petersen *et al.,* 1992). Local soil type will affect rates of erosion and infiltration of overland flows (*Rosgen,* 1993). In terms of buffer function, local topography can cause substantial differences in the distance of the riparian boundary from the stream. Steep slopes, especially when sparsely vegetated, will allow water to flow most quickly overland, decreasing opportunities for nutrient removal by both plants and soils as well as increasing erosion and subsequent sediment inputs. Boundaries can not be derived for buffer function in urban situations where polluted water enters the stream through storm and sanitary sewer outlets rather than flowing over or through the riparian area.

When the boundaries of riparian areas are based on their potential function as buffers, the distance from the stream will be highly variable depending on local soils and hydrological, topographical, and vegetation characteristics and should be determined for each unique stream reach.

The characteristics and extent of the riparian area interact in a variety of other ways with stream ecology. Most of the energy available to organisms low on stream food chains, such as invertebrates, is generated outside the banks of the stream by riparian vegetation (Hachmoeller *et al.,* 1991). The energy becomes available to the stream as carbon in the form of leaf litter and debris (Vannote *et al.,* 1980), and to a lesser extent, as dissolved nutrients (Adamus *et al.,* 1990). These inputs vary according to the spatial extent of riparian vegetation, its chemical characteristics depending on species, and the seasonal timing of inputs. Specifically, the litter of deciduous and coniferous species behave differently. Deciduous leaves tend to decompose quickly, and in temperate areas enter streams in seasonal peaks. Inputs of coniferous litter are much less seasonal, and the litter is slow to decompose. Where both are present there may be a balance between deciduous materials which are readily assimilated, and more evenly available coniferous litter (Gregory *et al.,* 1991). As suggested by the example in the Introduction, the area from which vegetation contributes substantial amounts of litter may be quite independent of hydrology, and the boundaries thus defined dependent on the size and type of adjacent vegetation.

Vegetation, especially forest, influences the availability of sunlight energy to streams. This has a direct effect on primary productivity by moderating the amount of light which gets to the water surface (Wiley *et al.,* 1990). Especially in small headwater streams, large forest trees can shade the stream (Barton *et al.,* 1985). Boundaries based on shading effects are necessarily relatively close to the stream bank, though dependent on vegetation size and structure. In systems where an abundance of nutrients are supplied to the stream by polluted runoff, high light levels allow blooms of nuisance algae to flourish. The direct heating effect of solar radiation to streams also interacts substantially with their ecology. Trout species, for example, require the high oxygen levels available in cold water. It has been shown that some fish and insects simply cannot survive, or at least barely subsist in warm water with otherwise adequate water quality. In a southern Ontario study, the only environmental variable which clearly distinguished trout and non-trout streams was weekly maximum temperature: at less than 22°C there were trout, above that, populations were marginal at best (Barton *et al.*, 1985).

MANAGEMENT OF THE RIPARIAN ECOTONE AND BOUNDARY

Management based on riparian boundaries set by the requirements of any single species is difficult to reconcile with an 'ecosystem approach' (Crombie, 1990). Neotropical migrant song birds are often used as representative species in a conservation context, however, a wide range of taxa including invertebrates, amphibians, reptiles, birds, and mammals are dependent on healthy and varied riparian vegetation. Boundaries based on habitat considerations are necessarily somewhat arbitrary boundaries of practical convenience. That is, different species may venture farther from the stream banks, or may go different distances to fulfill different seasonal or life cycle requirements. This has the result that, depending on the species under consideration, the effective boundaries of a riparian forest may differ substantially. Furthermore, all the characteristics and functions of riparian zones are dynamic in time and space, and this too has implications for decision makers, planners, and ecologists involved in restoration.

Pursuing a single species, or a single-goal approach, is tempting from a planning perspective since it is much easier to envision solutions to simple, rationally solvable tasks. For relevance to real conditions, attempts should be made to coordinate management efforts on a watershed scale to provide as much benefit as possible in the context of existing conditions at multiple scales and for many species.

ACKNOWLEDGMENTS

The comments and suggestions of Dr. B. G. Warner, Dr. G. Mulamoottil, and C. Hendrickson were all greatly appreciated. Research for this paper was supported by a Tricouncil EcoResearch grant.

REFERENCES

Adamus, P. R., Smith, R. D., and Muir, T., 1990. Manual for assessment of bottomland hardwood functions operational draft. Miscellaneous paper EL-90. US Army Engineer Waterways Experiment Station, Vicksburg, MS.

Armour, C. L., Duff, D. A., and Elmore, W., 1991. The effects of livestock grazing on riparian and stream ecosystems. *Fisheries* 16:7–11.

Baker, W. L., 1989. Classification of the riparian vegetation of the montane and subalpine zones of western Colorado. *Great Basin Naturalist* 49:214–228.

Barnes, W. J., 1983. Population dynamics of woody plants on a river island. *Canadian Journal of Botany* 63:647–655.

Barton, D. R., Taylor, W. D., and Biette, R.M., 1985. Dimensions of riparian buffer strips required to maintain trout habitat in southern Ontario streams. *North American Journal of Fisheries Management* 5:364–378.

Blom, C. W. P. M., Boegemann, G. M., Laan, P., and van der Sman, A. J. M., 1990. Adaptations to flooding in plants from river areas. *Aquatic Botany* 38:29–47.

Borman, F. H. and Likens, G. E., 1979. *Pattern and Process in a Forest Ecosystem.* Springer-Verlag, New York.

Collette, A., 1983. Effect of Spring Flooding on the Structure of Riparian Communities Along the Speed River in Southern Ontario, M.Sc. thesis, University of Guelph, Guelph, Ontario.

Correll, D. L., 1991. Human impact on the functioning of landscape boundaries. In: *Ecotones: The Role of Landscape Boundaries in the Management and Restoration of Changing Environments.* M. Holland, P. G. Risser, and R. J. Naiman, Eds. Chapman and Hall, New York, pp. 90–103.

Crombie, D., 1990. Regeneration: Toronto's waterfront and the sustainable city: final report. Minister of Supply and Services Canada, Toronto, Ontario.

Dillaha, T. A., Sherrard, J. H., Lee, D., Mostaghini, S., and Shanholtz, V. O., 1988. Evaluation of vegetative filter strips as best management practices for feed lots. *Journal of the Water Pollution Control Federation* 60:1231–1238.

Gosz, J. R., 1991. Fundamental ecological characteristics of landscape boundaries. In: *Ecotones: The Role of Landscape Boundaries in the Management and Restoration of Changing Environments.* M. Holland, P. G. Risser, and R. J. Naiman, Eds. Chapman and Hall, New York, pp. 8–30.

Gregory, S. V., Swanson, F. J., McKee, W. A., and Cummins, K. W., 1991. An ecosystem perspective of riparian zones. *BioScience* 41:540–551.

Hachmoeller, B., Matthews, R. A., and Brakke, D. F., 1991. Effects of riparian commmuntiy structure, sediment size, and water quality on the macroinvertebrate commmunities in a small, suburban stream. *Northwest Science* 65:125–132.

Harker, D., Evans, S., Evans, M., and Harker, K., 1993. Landscape Restoration Handbook. Lewis Publishers, Boca Raton, FL.

Hillbricht-Ilkowska, A. and Pieczynska, E. (eds.), 1993. Nutrient dynamics and retention in land/water ecotones of lowland, temperate lakes and rivers. Hydrobiologia 251:361.

Holland, M. M., 1988. SCOPE/MAB Technical consultants on landscape boundaries report of a SCOPE/MAB workshop on ecotones. *Biology International* Special Issue, 17:47–106.

Holland, M. M., Whigham, D. F., and Gopal, B., 1990. The characteristics of wetland ecotones. In *The Ecology and Management of Aquatic-Terrestrial Ecotones,* Man and the Biosphere Book Series. R. J. Naiman and H. Decamps, Eds. Parthenon Publishing, Carnforth, U.K., pp. 171–198.

Naiman, R. J., Decamps, H., and Pollock, M., 1993. The role of riparian corridors in maintaining regional biodiversity. *Ecological Applications* 3:209–212.

National Research Council (U.S.), 1992. *Restoration of Aquatic Ecosystems.* National Academy Press, Washington, D.C.

Peterjohn, W. T. and Correll, D. L., 1984. Nutrient dynamics in an agricultural watershed: observations on the role of a riparian forest. *Ecology* 65:1466–1475.

Petersen, R. C., Peterson, L. B.-M., and Lacoursiere, J., 1992. A building block model for stream restoration. In *River Conservation and Management.* P. J. Boon, P. Calow, and G. E. Petts, Eds. John Wiley & Sons, New York, pp. 293–309.

Phillips, J. D., 1990. Nonpoint source pollution control effectiveness of riparian forests along a coastal plain river. *Journal of Hydrology* 110:221–237.

Pickett, S. T. A., Kolasa, J., Armesto, J. J., and Collins, S. L., 1989. The ecological concept of disturbance and its expression at various hierarchical levels. *Oikos* 54:129–136.

Reily, P. W. and Johnson, W. C., 1981. The effects of altered hydrologic regime on tree growth along the Missouri River in North Dakota. *Canadian Journal of Botany* 60:2410–2423.

Risser, P. G., 1990. The ecological importance of land-water ecotones. In *The Ecology and Management of Aquatic-Terrestrial Ecotones.* R. J. Naiman and H. Decamps, Eds., Unesco and Parthenon Publishing, Carnforth, U.K., pp. 7–21.

Rosgen, D. L., 1993. Overview of rivers in the west. Riparian management: common threads and shared interests. A western regional conference on river management strategies. 1993 Feb 4–6, NM. Gen. Tech. Rep. RM-226. B. Tellman, H. J. Cortner, M. G. Wallace, L. F. DeBano, and R. H. Hamre, Eds., U.S.D.A. Forest Service, Rocky Mountain Forest and Range Experiment Station, Fort Collins, CO, pp. 8–15.

Sedell, J. R. and Swanson, F. J., 1990. Ecological characteristics of streams in old-growth forests of the Pacific Northwest. Symposium, Old Growth Forests...What are they? How do they work?, Toronto, January 20, 1990.

Stromberg, J. C. and Patten, D. T., 1990. Riparian vegetation instream flow requirements: a case study from a diverted stream in the Eastern Sierra Nevada, California, *Environmental Management* 14:185–194.

Tellman, B., Cortner, H. J., Wallace, M. G., DeBano, L. F., and Hamre, R. H., 1993. Riparian management: common threads and shared interest. A western regional conference on river management strategies., 1993 Feb. 4–6, NM. Gen. Tech. Rep. RM-226. U.S.D.A. Forest Service, Rocky Mountain Forest and Range Experiment Station, Fort Collins, CO.

van der Maarel, E., 1988. Vegetation dynamics: patterns in time and space. *Vegetatio* 77:7–9.

Vannote, R. L., Minshall, G. W., Cummines, K. W., Sedell, J. R., and Cushing, C. R., 1980. The river continuum concept. *Canadian Journal of Fisheries and Aquatic Sciences* 37:130–137.

Wiley, M. J., Osborne, L. L., and Larimore, R. W., 1990. Longitudinal structure of an agricultural prairie river system and its relationship to current stream ecosystem theory. *Canadian Journal of Fisheries and Aquatic Sciences* 47:373–384.

THE USE OF AVIAN FAUNA IN DELINEATING WETLANDS IN THE BALDWIN WETLAND COMPLEX, SOUTHERN ONTARIO

11

P. J. Harpley and R. J. Milne

ABSTRACT

An alternative view to the delineation of wetland boundaries based on the faunal component is presented. The behavior of breeding birds in the Baldwin wetland complex located in the northeast section of the Greater Toronto Area, Ontario indicates that wetland delineation should consider the faunal component along with the botanical, hydrological, and geomorphological criteria. Although a distinct boundary existed between the wetland and upland vegetation, when the ranges of resident birds were mapped the sharpness of this boundary was not as obvious. Reasons for this included (a) wetland avian species utilized the surrounding upland sites for various life functions including nesting and territorial defense, (b) species that nested in adjacent upland sites depended on the wetland for foraging, and (c) species from other landscape units in the fragmented landscape relied on this wetland as a food source. The implications of these factors on management of the wetland are discussed.

INTRODUCTION

An important issue in wetland management and classification is the delineation of its boundaries. Often such delineation relies on visually identifiable features such as the physical terrain, zonation of vegetation, and drainage patterns that reflect the change from one habitat to another (e.g., wetland to upland). However, boundaries based on other characteristics of these ecosystems can be much more difficult to establish from maps or by remote sensing. This is true for the faunal component, especially the larger and more mobile groups of animals such as birds and mammals. For instance, delineating specific boundaries based

1-56670-147-3/96/$0.00+$.50

on wetland species is complicated by the fact that an animal may rely on a wetland for only a portion of its life history. These species move between habitats for their sustenance such as foraging and nesting and even the defense of territorial space. Therefore, their survival is dependent on a mosaic of habitats or on the landscape. Unfortunately, for many species the details of their life histories including daily movements and habitat use have not been studied extensively, especially behavioral responses to the pattern of the landscape.

Research on delineating the territory of species and the area required to maintain viable populations has focused on various characteristics of life histories of animals with a view to establish guidelines for minimal areas for groups of species in specific habitats. The size of forests for successful breeding has been estimated for some area-sensitive species (Robbins et al., 1989) and the spatial limits of feeding ranges has been determined (Schoener, 1968). Other authors have estimated minimal landscape unit sizes and their research has been directed at specific, homogeneous habitats, especially viable sizes of woodlots for forest interior migrants (Robbins et al., 1989; Nudds, 1993). There has been limited research to determine the territorial areas of species in wetland habitats or species that occur in fragmented landscapes of wetlands and upland forests in moderately to heavily disturbed environments. In certain instances, conserving habitats as small as 10 ha may be necessary to protect populations of some species in fragmented forested landscapes, as for example in the Oak Ridges moraine, Ontario (Cranmer-Byng, 1994). There has also been little work to delineate the combined spatial requirements for breeding success with other functions such as foraging, territorial defense, mate attraction, protection, and predator alert (Crocker-Bedford, 1990).

In this chapter, avian groups from the Baldwin wetland complex in southern Ontario have been identified based on their dependence or independence to specific wetland habitats. Representative species from these groups have been selected and their territories during the breeding season have been mapped to show the complexity of their spatial patterns. These species included both rare and common birds dependent on the wetland complex, as well as those that were not dependent on this habitat but were important in the functioning of the wetlands. The preliminary results of this work are discussed with respect to the problems and importance of considering the faunal component in the delineation of wetland boundaries. This research is part of a long-term research program on the effects of human disturbance on avian diversity and abundance at sites in the Greater Toronto Area (GTA) (Milne and Harpley, 1994).

The GTA is experiencing increasing pressures from an expanding population base and from commercial and industrial development which is constantly increasing the fragmentation of the natural landscapes and the loss of critical habitats, particularly wetland complexes. With these pressures there is an increasing demand for the proper management of wetlands and surrounding lands. Unfortunately, there is little information available in this region as to how they function in an integrated manner.

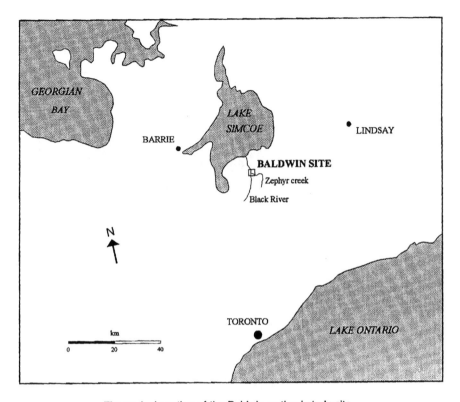

Figure 1 Location of the Baldwin wetland study site.

SITE DESCRIPTION

The Baldwin wetland complex is located in the town of Georgina, south of Lake Simcoe in the northern region of the GTA. The wetland complex is located at the junction of two streams that flow into an open water wetland. The complex includes stretches of the Black River and Zephyr Creek that drain from the north slope of the Oak Ridges moraine through the southern end of the Lake Simcoe drainage basin (Figure 1). The streams are characterized by meandering channels flowing through beds of *Typha latifolia*. The surrounding uplands are covered by a combination of mature forests and successional fields and forests.

The relatively undisturbed nature of the site makes this area useful as a natural benchmark for monitoring any future changes for the region. The wetland complex is particularly unusual for the GTA in that it is relatively inaccessible and has virtually no development on the shores of the rivers and much of the area bordering the wetland. However, some modest development pressures from agriculture and urban influences are increasing the fragmentation of the landscape. There are sections of the forest interior dissected in several locations by agricultural fields, residential developments, and powerline corridors. A general pattern of the landscape and fragmented habitats is presented in Figure 2.

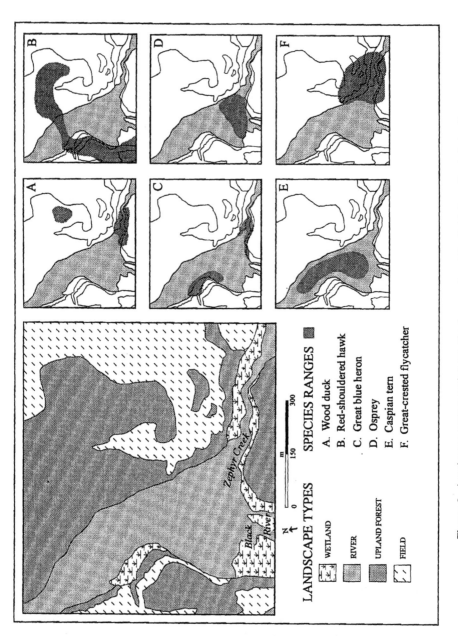

Figure 2   Landscape pattern and range of avian species at the Baldwin wetland study site.

METHODS

The status of the breeding birds for the Baldwin wetland complex was determined over three seasons, 1991 to 1993, commencing in late May and extending until the middle of August. Population size and breeding success were recorded for each species as well as behavior and spatial utilization of the landscape during the 1992 and 1993 seasons. This included identifying the position and movement of birds within and between landscape units. For example, individuals were found nesting in the wetland, perching on floodplain trees for territorial defense, and foraging in upland forests.

Unlike many bird censuses where a limited time, generally several hours, is allotted to specific plots, in this study field surveys were conducted two or three times a week from late May to late August in 1992 and 1993 to obtain a complete pattern of species activity. The sites were sampled primarily in the early morning hours from 6:00 until 11:00 a.m. when the majority of species were most active. However, every fourth sample period was scheduled from 2:00 until 9:00 p.m., to include those species more active at this time, such as the common nighthawk (*Chordeiles minor*). Approximately equal field hours were spent on foot and by canoe. Because of the fragmented nature of the site it was decided to sample the entire site rather than to limit the research to sample plots. This ensured that all unique and indicator species were identified and that the smallest fragmented habitats were sampled.

RESULTS

The results include a classification of the rare and common avian populations associated with the wetlands and an initial mapping of representative wetland species based on their behavior, spatial patterns, and utilization of the landscape. Populations are presented as the total number of individuals recorded during the entire study period or breeding season and averaged for the two-year period, 1992–1993. Results of inventories completed during the period 1991 to 1993 have been summarized in other reports (Harpley and Pollock, 1991; Joyce *et al.*, 1993; Milne and Harpley, 1994). The range or distance that the individuals travelled during the observation periods have also been estimated for some species. In many cases these distances represent a minimum distance and the actual range is much greater.

Species Composition

At the Baldwin wetland complex several species recognized as rare or vulnerable at various administrative levels were identified. Table 1 shows population numbers for those species considered rare or endangered at the national (COSEWIC, 1993), provincial (Austen *et al.*, 1994), and regional level (Ecologistics, 1982). At the Baldwin study site there was breeding evidence and historical records of the red-shouldered hawk (*Buteo lineatus*) and least bittern (*Ixobrychus exilis*) and

Table 1   Populations of Rare Bird Species at the Baldwin Wetland Complex

Common name	Rarity status	Scientific name	Minimum distance (km)[c]	Population
Hooded merganser	Local[a]	*Lophodytes cucullatus*	—	1
Caspian tern	National[b]	*Sterna caspia*	30.0	2
Least bittern	National	*Ixobrychus exilis*	0.5	1
Red-shouldered hawk	National	*Buteo lineatus*	1.0	1
Sedge wren	Local	*Cistothorus platensis*	0.2	2
Eastern bluebird	National	*Sialia sialis*	—	1

[a] Identified as rare by the Lake Simcoe Regional Conservation Authority (Ecologistics, 1982).

[b] Identified as vulnerable by the Committee on the Status of Endangered Wildlife in Canada (COSEWIC, 1993).

[c] Minimum distances were not determined for several species.

sightings of eastern bluebird (*Sialia sialis*), Cooper's hawk (*Accipiter cooperii*), and Caspian tern (*Sterna caspia*), all listed as vulnerable at the national level (COSEWIC, 1993). There were also sightings and historical records of sedge wren (*Cistothorus platensis*) and hooded merganser (*Lophodytes cucullatus*), recognized as rare by the LSRCA (Ecologistics, 1982; Harpley and Pollock, 1991).

It should be noted that this site has been identified as an important habitat for a number of rare species in the GTA. Of equal importance was occurrence of the common species with high diversity and populations that utilized the wetland for part of their life history. These were all an integral component of the wetland complex. For example, flycatchers and swallows accounted for a large percentage of the total populations in the study area. There were several species of flycatchers including the least flycatcher (*Empidonax minimus*), eastern phoebe (*Sayornis phoebe*), eastern kingbird (*Tyrannus tyrannus*), great crested flycatcher (*Myiarchus crinitus*), and eastern wood-pewee (*Contopus virens*). Swallow species included the tree swallow (*Tachycineta bicolor*), barn swallow (*Hirundo rustica*), bank swallow (*Riparia riparia*), and northern rough-winged swallow (*Stelgidopteryx serripennis*).

Spatial Patterns and Landscape Use

The species were classified based on their use of the wetlands and surrounding habitats to simplify the complexity of the spatial patterns of species within this area. The primary classification categorized the species as either landscape dependent or independent. 'Landscape dependent' referred to those that were dependent either on a specific habitat or combination of habitats for maintaining a population. 'Landscape independent' was applied to species that were not dependent on specific habitats or combinations of habitats, but were important members of the Baldwin wetland complex. The secondary classification identified the dependent habitat where the species were commonly found and other habitats that the species

Table 2   Populations for Wetland Dependent-Upland Associated Types at the Baldwin Wetland Complex

Common name	Scientific name	Minimum distance (km)	Population
Wood duck	Aix sponsa	0.5	7
Belted kingfisher	Ceryle alcyon	0.6	3
Red-winged blackbird	Agelaius phoeniceus	0.1–0.2	56

were associated with for a portion of their life history such as nesting, foraging, and territorial defense.

Landscape Dependent Species

Wetland Dependent-Upland Associated

'Wetland dependent' refers to species that were commonly recognized as totally or partially dependent on wetlands and required this habitat for most of their life history (OMNR, 1991; FON, 1992). These species were also associated with the undisturbed forested areas surrounding the wetland. Examples of these species are provided in Table 2. The wood duck (Aix sponsa) used the wetland for feeding and protection but nested in cavities in trees that bordered the adjacent stream corridor. Its range is depicted in Figure 2A. There were two locations including the riparian corridor of Zephyr Creek, especially near the mouth of the stream, where the tree cover on both sides of the channel was greater. These ducks were also observed feeding in small wet depressions in the forest interior of the upland forest to the north of the stream. In this area of the forest the drainage was poor and the forest was dominated by dense stands of eastern white cedar (Thuja occidentalis) that provided a protective cover.

The red-winged blackbird (Agelaius phoeniceus) was also included in this group. The bird nested and foraged primarily in the wetlands, but the males used the surrounding larger trees as an important perch to defend their territory, attract mates, and warn of predators. The greatest number of birds were observed in the Black River corridor, which had the largest area of marsh. Smaller populations occurred along Zephyr Creek and in a small cattail marsh on the west side of the lake, north of the mouth of the Black River. Preliminary observations indicated that there may be a relationship between the location and height of perch trees and nest location and density. Within the Black River corridor, there was a greater concentration of nests near the edges of the corridor. This spatial pattern of nests suggested that the perches along the shore may be an important factor in nest location although it could be a function of other factors such as emergent vegetation communities.

Upland Dependent-Wetland Associated

Upland dependent-wetland associated species required a combination of landscape units including upland forests and fields and wetland corridors. This included large stretches of forest interior that provided nesting sites and riparian

Table 3 **Populations for Upland Dependent-Wetland Associated Types at the Baldwin Wetland Complex**

Common name	Scientific name	Minimum distance (km)	Population
Red-shouldered hawk	*Buteo lineatus*	1.0	1
Northern harrier	*Circus cyaneus*	—	2
Hairy woodpecker	*Picoides villosus*	—	2
Northern rough-winged swallow	*Stelgidopteryx serripennis*	—	3
Tree swallow	*Tachycineta bicolor*	0.8	20
Yellow warbler	*Dendroica petechia*	0.1	8
White-throated sparrow	*Zonotrichia albicollis*	0.1–0.2	4

Table 4 **Populations for Wetland/Pond Dependent-Landscape Associated Types at the Baldwin Wetland Complex**

Common name	Scientific name	Minimum distance (km)	Population
Osprey	*Pandion haliaetus*	7.0	2
Great blue heron	*Ardea herodias*	5.0	6
Green-backed heron	*Butorides striatus*	—	3
Caspian tern	*Sterna caspia*	10.0	2

corridors for foraging and protection. This landscape pattern attracted several raptor species including the red-shouldered hawk (Table 3). Northern harrier (*Circus cyaneus*) were also found where there was a combination of abandoned fields and wetlands. The presence of the red-shouldered hawk at the study area supported the findings of Bosakowski *et al.* (1992), who observed that this species nested in landscapes that included larger wetlands and stream channels in forests that had a greater proportion of coniferous and mixed forest, a pattern typical of this site (Figure 2B). Their findings also suggested that this species was area sensitive, and a reduction in the amount of either of these habitats would eliminate the bird from the landscape.

Other species associated with this landscape pattern included the tree swallow, northern rough-winged swallow, hairy woodpecker (*Picoides villosus*), yellow warbler (*Dendroica petechia*), and white-throated sparrow (*Zonotrichia albicollis*). The tree swallow and hairy woodpecker are cavity nesters and dependent on tree hollows in the upland forest for nest sites, but are often associated with wetlands for part of their life histories. The tree swallow was an important member of the wetland complex since this species occurred in high numbers and foraged predominantly in the wetlands.

Wetland Dependent-Wetland Associated

Species in this group included those that relied on the lake and wetland areas of the Baldwin wetland complex yet were not residents of the area (Table 4). They nested elsewhere in the region in other wetlands and travelled several kilometers to the Baldwin complex to feed in the wetland habitats. They used the study area for specific time periods which could be measured on a daily and seasonal basis. Therefore, they were associated with the wetland habitats at Baldwin, but more generally, were dependent on the pattern of wetlands and small lakes throughout the southeast portion of the Lake Simcoe drainage basin.

The great blue heron (*Ardea herodias*) was most often observed during the day, especially in the mornings feeding in cattail stands. It would spend a portion of the day throughout the breeding season feeding in this section of the wetland although it nested elsewhere in the region. The usual location of its activities was near the mouth of Zephyr Creek and on the west shore of the pond (Figure 2C). Gibbs *et al.* (1991) noted that this species was more common in wetlands with extensive beds of emergent vegetation and less open water, which supports the choice of forage sites observed in this study. Gibbs *et al.* (1991) also found that heron abundance decreased with increased isolation of the wetland. This suggests that the pattern and density of wetlands in the region would have an impact on the presence of this species. The Baldwin study site favors the presence of this species since it is located in an area of the GTA where there is a high proportion of relatively undisturbed wetlands linked by natural corridors of wetlands and forests.

Osprey (*Pandion haliaetus*) was also a common visitor to the wetlands and pond throughout the breeding season. The bird would typically arrive in midmorning and hunt most often at the confluence of the streams where the Zephyr creek meets the Black River (Figure 2D). This site was an important source for food although the birds nested outside of the study area. This species also prefers wetlands in close proximity to one another as this situation provides a number of alternate foraging sites within close range of the nest sites (Gibbs *et al.*, 1991).

Caspian tern has been included in this category, although it did not nest in adjacent wetland. It used the pond for feeding in midsummer near the end of the breeding season (Figure 2E). During this part of the season, it appears that individuals dispersed to small lakes and wetlands in the region to feed, returning to the colony at night. The only breeding location for this vulnerable species in the region was located approximately 10 km to the north on Lake Simcoe near Georgina Island (Austen *et al.*, 1994). The birds used the Baldwin study site only to feed and likely returned to Lake Simcoe later in the day. The populations of the colony may be increased in midsummer as birds from northern colonies on the Great Lakes move south. Tozer and Richards (1974) reported that there was an influx of terns on Lake Scugog, situated to the east of the study site, beginning in the first half of July. Like other species in this group, it was apparent that the availability of alternate food sources in close proximity to the nesting grounds was an important factor in the feeding patterns of the terns.

Landscape Independent Species

Upland Associated-Wetland Independent

Species in this group did not necessarily require wetlands and were found in a variety of habitats. However, at this study site these species used the wetland corridors as a prime foraging area and the open space of the corridors for display

Table 5   Populations for Upland Associated-Wetland Independent Types at the Baldwin Wetland Complex

Common name	Scientific name	Minimum distance (km)	Population
Northern flicker	Colaptes auratus	0.7	10
Eastern kingbird	Tyrannus tyrannus	0.3	14
Great crested flycatcher	Myiarchus crinitus	0.4	12
Eastern wood-pewee	Contopus virens	0.3	10
Least flycatcher	Empidonax minimus	0.2	8
Bank swallow	Riparia riparia	—	15
Barn swallow	Hirundo rustica	0.8	15
Cedar waxwing	Bombycilla cedrorum	—	15
Warbling vireo	Vireo gilvus	0.2	5
Song sparrow	Melospiza melodia	0.1–0.2	22

and defense, although their nests were located in the surrounding forests and open fields. Often individuals were found in the large trees that bordered the wetlands. These perch locations were used by many species since they provided a good vantage point for mate attraction and protection of territory from other individuals as well as from predators.

A large number of species was included in this group (Table 5). Some important representative groups were the flycatchers including the great crested flycatcher, least flycatcher, eastern wood-peewee, eastern phoebe, and eastern kingbird, and the swallows including the bank swallow and barn swallow. Some other common species included warbling vireo (*Vireo gilvus*), northern flicker (*Colaptes auratus*), song sparrow (*Melospiza melodia*), and cedar waxwing (*Bombycilla cedrorum*). As mentioned earlier, these species constitute a significant portion of the total avian population for this landscape. Average population numbers are presented in Table 5. Some of the greatest numbers are represented by the swallows and flycatchers. If the tree swallow and northern rough-winged swallow (Table 3) are included with the barn swallow and bank swallow, the population averaged approximately 50 individuals a season. Similarly, the flycatchers accounted for 45 individuals.

A representative range for a pair of great crested flycatchers has been mapped in Figure 2F. This map shows the overlap of the nest and forage areas, which included a significant area in the upland forest as well as the wetland corridor. A similar pattern was found for species that were dependent on the upland fields that also lined the wetland corridor at the east end of the site. The abandoned fields and pastures hosted many species including the eastern kingbird and yellow warbler that were observed using the wetlands for foraging and display.

DISCUSSION

The territorial ranges of many species found in the Baldwin wetland complex were centered in the wetlands but included significant portions of the upland forest and/or field. The great crested flycatcher, representative of many of the landscape independent species, had a range that was greater than the limits of the wetland habitat. Similarly, many species that were in groups that were landscape dependent covered areas that extended beyond the wetland. Raptors including the red-shouldered hawk had a territory that included nest sites in the upland forest within a close proximity to wetlands and a large foraging range that included the combined natural and agricultural landscapes typical of the northern GTA. Wetland species such as the wood duck nested within or near the wetland, but also fed in flooded depressions within the forest. Alternatively, species such as the great blue heron and osprey nested in adjacent wetlands but fed within the Baldwin wetland complex.

If the faunal component is used to define wetland boundaries, then this would extend the boundaries of the wetlands far beyond the wetland plant communities at this site. Species such as the wood duck would extend the 'wetland' to include areas that are not traditionally included within wetland boundaries. It is recognized that because of variations between species ranges and the large number of species associated with wetland/forest habitat complexes, the use of bird territories in delineating wetlands could be a complicated procedure. This could be simplified by focusing on representative or indicator groups, for instance those that are dependent on wetlands and associated with upland forests. Wildlife management policies are often put in place to protect the habitat and viable populations of threatened or game species (Forbes, 1993). One of the biological aspects considered in the wetlands policy statement is wildlife disturbance zones specifically for endangered raptors and migratory birds (OMNR, 1992). These species may require zones of undisturbed nesting habitat which may be beyond wetland boundaries. If the faunal component is not considered in the mapping of a wetland, critical habitats from the surrounding landscape essential for the existence of some wetland species will not be included, and subsequent management decisions based on these delineations may be detrimental to their populations.

Although focusing on individual species and their nesting requirements is necessary to ensure that they are not at risk because of declining habitat area and quality, it should not be the only criteria considered when developing wildlife and habitat management strategies in wetlands. A critical component of wetland management is the maintenance of wetland functions (e.g., OMNR, 1992). Bird populations are an important component of the wetland ecosystem as an integral member of the food chain and less directly as a control of stream characteristics such as water quality and nutrient uptake. These functions will be most affected by birds that are present in large numbers. Consequently, wetland dependent and wetland associated species which form a significant portion of the avian population will have a greater impact on the wetland than small colonies of threatened species.

Wetland boundaries should also include the impact of species not dependent on wetlands but in specific cases use this habitat. For instance, flycatchers and swallows are dominant members of the food chain at the Baldwin wetland complex. Species such as the great crested flycatcher, eastern phoebe, and eastern kingbird are some of the main insect feeders in this wetland, although they predominantly nest in the upland forests and fields. These species will have a significant impact on the food chain, both as predator and prey. As an important component in the functioning of these wetlands, they should be included in defining the limits of the wetland.

The spatial patterns of the birds examined in this chapter suggest that if the surrounding forest is further fragmented then many of these species will be lost from this landscape. In an urbanizing landscape where wetlands are preserved but surrounding forest areas are sacrificed this can have detrimental long-term impacts on many avian species. Disturbance or loss of lands outside the traditional boundary of the wetland will reduce population levels while at the same time disrupt the food chain. To prevent this from taking place there has to be greater recognition of the importance of avian species to the wetland complex and their role in processes of this ecosystem. This can be accomplished by reexamining the criteria and methods used in classifying and delineating wetlands, by identifying areas that are necessary for the maintenance of avian populations and the wetland, but are traditionally considered beyond the boundaries of the wetland.

CONCLUSIONS AND RECOMMENDATIONS

As a response to the need to refine the criteria for the delineation of wetlands and the definition of wetland functions (e.g., OMNR, 1992) a few conclusions and recommendations with respect to birds are presented based on our research at the Baldwin wetland complex. Our work suggests that more attention needs to be paid to the avifauna in the delineation of wetland boundaries.

Although some delineation of wetlands are easily made in the field, the spatial distribution of some components of the landscape are not fixed and are perhaps less obvious, such as the amphibians, birds, and mammals. Many species travel between different landscape units and are dependent on each of the habitats for different life functions. This creates ranges that do not conform with patterns of terrain, vegetation, or hydrology and, ultimately, traditional wetland boundaries.

Boundaries should recognize that many functions of specific habitats or individual land units of the landscape, especially wetlands, are strongly linked to adjacent units such as upland forests and agricultural fields. The wetland should be viewed as one part of a much larger landscape mosaic. Different habitats of the landscape can serve as separate sites for nesting, foraging, and protection for many species, and it is the entire landscape mosaic that is often essential for these species survival. For example, a diverse landscape pattern is critical for such threatened species as the red-shouldered hawk.

Often, the importance of the faunal component to the health and stability of the wetland is overlooked. More numerous species, such as swallows and fly-catchers, including those that are not dependent on specific habitats, play a much greater role in the functioning of the wetland than threatened species. These groups of species are critical in the wetland food chain, especially in controlling the insect populations and as a food source for other birds higher in the food chain.

Therefore, life histories of birds and dependence on landscape pattern should be studied more vigorously and included in land use planning decisions.

Greater attention has to be paid to developing proper methods to inventory and monitor bird populations. In many environmental assessments the focus is on the nesting habitat to determine if the species is a breeder within a designated site. However, many species use the wetland for other functions such as a food source, territorial defense, protection, and mate attraction. These functions should be given similar weight when assessing the site. Similarly, inventories should not be restricted to species during breeding periods. Often, wetlands are important throughout the year for functions such as postbreeding dispersal and migration staging areas. Assessment should also not be restricted to single-day surveys. These are inadequate because a single time period will overlook species whose use of the wetland is not on a daily basis or always at a specific time, for example early morning. Variations in weather conditions such as precipitation and temperature can also disrupt bird activity and observation records. Similarly, variations in presence can occur from year to year, especially with less common species. Monitoring over a number of years is required to understand the dynamic nature of sites.

ACKNOWLEDGMENTS

This research was supported by funding from the Environmental Youth Corps Program, Ontario Ministry of Natural Resources, and the James L. Baillie Memorial Fund, Long Point Bird Observatory. The authors wish to thank the South Lake Simcoe Naturalists for assisting with the operation of this project. E. Joyce and C. Rothfels assisted with collecting and compiling the data. The following individuals assisted in the field work: N. Evans, L. Jessen, J. Maddin, R. Moore, P. Pollock, C. White, R. Wolske, and S. Zourdoumis. B. Morber and D. Harpley provided technical advice and management support, respectively.

REFERENCES

Austen, M. J. W., Cadman, M. D., and James, R. D., 1994. *Ontario Birds at Risk:Status and Conservation Needs.* Federation of Ontario Naturalists, Don Mills, Ontario and Long Point Bird Observatory, Port Rowan, Ontario. 165p.

Bosakowski, T., Smith, D. G., and Speiser, R., 1992. Status, nesting density, and macrohabitat selection of Red-shouldered hawks in northern New Jersey. *Wilson Bulletin* 104:434–446.

COSEWIC, 1993. Canadian species at risk. Committee on the Status of Endangered Wildlife in Canada, Ottawa, Ontario. 11p.

Cranmer-Byng, M., 1994. Is draft strategy for Moraine strong enough? *Seasons* 34:12.

Crocker-Bedford, D. C., 1990. Goshawk reproduction and forest management. *Wildlife Society Bulletin* 18:262–269.

Ecologistics Limited, 1982. Environmentally significant areas study. Lake Simcoe Regional Conservation Authority, Newmarket, Ontario. 352p.

Federation of Ontario Naturalists, 1992. *Birds of Southern Ontario's Wetlands.* Federation of Ontario Naturalists, Don Mills, Ontario. 4p.

Forbes, G. J., 1993. The preservation of genes, populations and habitat specialists in natural heritage areas, in *Size and Integrity Standards for Natural Heritage Areas in Ontario*, Poser, S. F., Crins, W. J., and Beechey, T. J., Eds., Parks and Natural Heritage Policy Branch, Ontario Ministry of Natural Resources, Huntsville, Ontario, pp. 34–46.

Gibbs, J. P., Longcore, J. R., McAuley, D. G., and Ringelman, J. K., 1991. Use of wetland habitats by selected nongame water birds in Maine. *Fish & Wildlife Research - U.S. Fish & Wildlife Service* 9:1–57.

Harpley, P. J. and Pollock, P., 1991. Breeding bird census of the sensitive Black River/ Zephyr Creek wetland/wildland complex at the Baldwin Pond in the Town of Georgina. South Lake Simcoe Naturalists, Georgina, Ontario, Technical Paper 1:1–4.

Joyce, E., Rothfels, C., Harpley, P. J., and Milne, R. J., 1993. Habitat diversity and life histories of vulnerable avian species, Greater Toronto Area. South Lake Simcoe Naturalists, Georgina, Ontario, Technical Paper 3:1–12.

Milne R. J. and Harpley, P. J., 1994. Landscape ecology, avian information and the rehabilitation of wildland complexes in the Greater Toronto Area, in *Proceedings of 1993 Canadian Land Reclamation Association Annual Meeting*, Lindsay, Ontario, August 11th–13th, Canadian Land Reclamation Association, Toronto, pp. 48–56.

Nudds, T. D., 1993. Use of estimated "pristine" species-area relations to generate indices of conservation value for nature reserves, in *Size and Integrity Standards for Natural Heritage Areas in Ontario*, Poser, S. F., Crins, W. J., and Beechey, T. J., Eds., Parks and Natural Heritage Policy Branch, Ontario Ministry of Natural Resources, Huntsville, pp. 25–33.

Ontario Ministry of Natural Resources., 1991. *Looking Ahead: A Wild Life Strategy for Ontario*. Ontario Wildlife Working Group, Queen's Printer for Ontario, Toronto, Ontario. 172p.

Ontario Ministry of Natural Resources. 1992. *Manual of Implementation Guidelines for the Wetlands Policy Statement*. Queen's Printer for Ontario, Toronto, Ontario. 108p.

Robbins, C. S., Dawson, D. K., and Dowell, B. A., 1989. Habitat area requirements of breeding forest birds of the middle Atlantic states. *Wildlife Monographs* 103:1–34.

Schoener, T. W., 1968. Sizes of feeding territories among birds. *Ecology* 49:123–141.

Tozer, R. G., and Richards, J. M., 1974. *Birds of the Oshawa-Lake Scugog Region, Ontario*. Alger Press, Oshawa. 383p.

TEMPORAL DELINEATION OF WETLANDS ON GULL POINT, PRESQUE ISLE, PENNSYLVANIA

P. S. Botts and R. Donn

ABSTRACT

Presque Isle is a recurved sand spit on the south shore of Lake Erie near Erie, Pennsylvania. Sand eroded from lake side beaches and transported to the distal end of the peninsula (Gull Point) has resulted in the extension of Gull Point and concurrent formation of shallow wetlands. The objectives of this chapter are to (1) demonstrate the use of aerial photographs to document the temporal and spatial history of wetlands on Gull Point, (2) analyze the spatial and temporal dynamics of the wetlands, and (3) create a spatial and temporal framework within which hypotheses regarding the population dynamics of the biota in the wetlands on Gull Point could be generated and tested. A series of aerial photographs and image analysis techniques were used to determine the ages and initial shapes of wetlands on Gull Point, subsequent changes in the spatial configuration of the wetlands as they became vegetated, and the probability that a newly formed wetland would survive to become an established wetland. Of 17 wetlands established on Gull Point between 1946 and 1993, 15 were formed after 1973. Most formed on the bay side of Gull Point as elongated lagoon wetlands that became isolated from the bay by sand deposition. As lagoon wetlands aged, they became subdivided into wetland "systems" in which encroachment by *Phragmites* reduced the area of open water and isolated individual wetlands. Wetlands formed on the lake side of Gull Point as beach wetlands behind sand berms created by wave action during storms. Beach wetlands tended to persist for less than five years prior to destruction by wave action.

1-56670-147-3/96/$0.00+$.50
© 1996 by CRC Press, Inc.

INTRODUCTION

Wetlands are spatially and temporally dynamic habitats (Mitsch and Gosselink, 1993) and the spatial boundaries of wetlands may change over time (Botts and McCoy, 1994). In coastal wetlands, geological and hydrological processes associated with erosion and deposition can create, alter, or destroy wetlands over remarkably short periods of time, changing greatly the biological processes occurring in the habitat. Biologists often expend a great deal of effort trying to obtain a temporal record of the spatial boundaries (the "history") of coastal wetlands (Lyon, 1993) because the history of patches in a heterogeneous habitat may provide a template from which biological dynamics can be studied or predicted (Allen and Starr, 1982; Pickett and White, 1985; Kolasa and Rollo, 1991).

Unfortunately, determination of the history of wetlands often must be based upon indirect evidence. Coastal wetlands on Lake Erie have been dated using sediment core samples (Ellenberger *et al.*, 1973; Reeder and Eisner, 1994), tree ring data (Jennings, 1909; Kormandy, 1969), vegetational analysis for seral stage (Jennings, 1909), and archival maps and anecdotal evidence. While these methods often are the only ones available, they can lead to conclusions that may not be accurate. Furthermore, they provide only limited spatial and temporal resolution of the true history of the wetlands in relation to the upland habitat that surrounds them.

Aerial photography has become an important tool for site assessment, environmental regulation, and management in a number of habitats (Lyon, 1987, 1993), although few such assessments are published in the scientific literature. One frequently overlooked use for aerial photography is determination of a detailed history of a habitat (Lyon and Drobney, 1984; Lyon and Greene, 1992).

In this chapter, the authors take the position that the spatial changes in wetland boundaries should, whenever possible, be studied in a temporal context. This approach is illustrated using aerial photographs and image analysis as tools for determining the history of wetland boundaries. The objectives of this chapter are to (1) demonstrate the use of aerial photographs to document the temporal and spatial history of wetlands, (2) analyze the spatial and temporal dynamics of the wetlands on Gull Point, and (3) create a spatial and temporal framework within which hypotheses regarding the population dynamics of the biota in the wetlands on Gull Point could be generated and tested.

METHODS

Study Site

The authors were fortunate to work in Presque Isle State Park, Pennsylvania, a habitat for which a particularly complete set of historical photographs exists. A series of 31 aerial photographs of Gull Point, the distal end of Presque Isle State Park, that were taken between 1946 and 1993 were examined. A combination of visual inspection and image analysis was used to determine the ages, initial boundaries, subsequent spatial changes, and longevity of all wetlands on Gull Point.

Presque Isle State Park is a 12-km recurved sand spit that extends into Lake Erie from Erie, Pennsylvania. The peninsula is composed of sand and gravel eroded from glacial deposits from the southern shore of Lake Erie by littoral currents and deposited to form the sand spit. The peninsula has been migrating along the shore of the lake toward the north-east on a platform of sand and gravel that was deposited by receding glaciers, but it has reached the end of the platform and is unlikely to move farther (Pennsylvania DER, 1993).

Gull Point is a compound sand spit located at the distal end of Presque Isle (Figure 1). Wetlands form when water becomes trapped behind sand bars or berms at the end of the peninsula (Jennings, 1909; Kormandy, 1969). Once a wetland forms, the trajectory of its subsequent development may be uncertain due to a combination of continuing impacts from erosional and depositional processes and the development of emergent vegetation. As a result, Gull Point wetlands are highly dynamic.

Aerial Photographs

The U.S. Army Corps of Engineers has been taking aerial photographs of Presque Isle since 1946. In the 1970s, the Corps began archiving a series of photographs intended to document the effects of various beach renourishment and breakwall projects on the spatial configuration of the peninsula. Photographs between 1946 and 1977 were taken at irregular intervals and at varied elevations, but photographs since 1977 have been taken at least annually from directly above the peninsula and produced at a scale of 1:4800.

Image Analysis

Aerial photographs of Gull Point were obtained from the Park Administration and digitized using a Targa M8 board and Bioquant OS/2 image analysis software. Images were captured using a Javelin CCTV black and white video camera fitted with a 16/1.6 lens. Photographs were digitized in such a way that landmarks in succeeding pictures were aligned in the same manner with regard to a north/south axis (Figure 2). Wetlands were measured using thresholding and automatic measurement based on gray-scale differences in the photographs. In those cases where the computer was unable to distinguish two shades of gray, wetlands were outlined and thresholded manually using a mouse, and then were measured automatically. Because the vegetation extended only a minimal distance into the wetlands, the open water area of each was easily measured.

The age and history of each wetland were determined by examining the photographs in chronological order beginning with the present-day location and extending back in time until the origin of each wetland could be discerned. For each wetland, its origin (lagoon formation or beach pond formation), year during which it was isolated from the lake, year of any subsequent alteration in the form (e.g., separation into different wetlands), and year during which it disappeared (when relevant) were recorded.

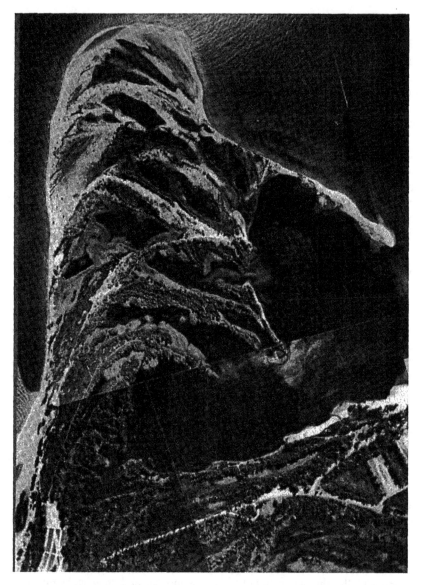

Figure 1   Digitized photograph of Gull Point, Presque Isle State Park, Erie, PA taken on July 16, 1993. The original photograph was taken in color and printed at a scale of 1:4800 from directly overhead. The dimensions of the area included in this photograph are 1900 m by 1356 m.

Data Analysis

Early in the analysis of the data, it became clear that, even when the edge of a wetland was readily discernable on a photograph, establishing the threshold values needed for the computer to distinguish the edge was a semisubjective procedure. Differences in the angle at which sunlight reflected from the water in

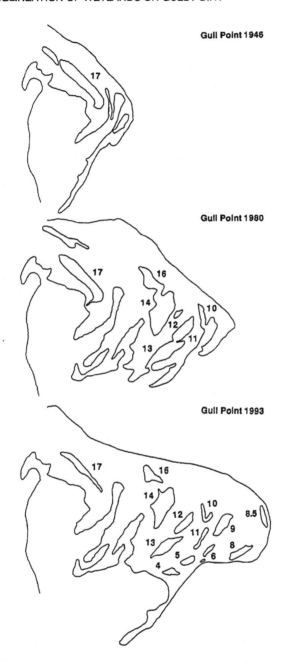

Figure 2   A series of drawings made from selected aerial photographs of Gull Point, Presque Isle State Park, Pennsylvania. All photographs were taken at the same scale and the drawings are oriented so that landmarks are aligned on the same x,y-coordinates. Numbers refer to wetlands present in the 1993 photograph, and the wetlands retain the same number in each drawing. Thus, Wetland 17 was present in 1946, 1980, and 1993. Wetlands 10 and 11 were present in 1993, and both were present in 1980 as parts of Lagoon 10,11; they were not present in 1946.

the wetlands and wide bands of shallow or sediment-laden water within wetlands and lagoons created nonuniform colors in areas of open water (Figure 1). In several cases, the computer could not distinguish between vegetation near the edges of wetlands and water. In such cases, it was necessary to draw wetland boundaries manually with the computer.

To determine the degree of subjective error introduced during thresholding and/or manual drawing, seven wetlands from the 1993 photograph were chosen for closer statistical scrutiny. Five naive viewers were asked to set thresholds and/or manually draw boundaries and measure the wetlands independent of each other. The data were used to determine the range of variances that might be expected around our measurements.

The area of Gull Point, determined annually between 1977 and 1993 (the dates with photographs taken from the same angle at the same scale), was plotted against time, and a least-squares regression was calculated to determine the rate at which Gull Point was growing. Considerable debate regarding the influence of beach renourishment on the rate of growth/wetland area of Gull Point prompted the calculation of a second regression that included the area of Gull Point in 1946 (a date with a photograph taken at the same angle and scale as the more recent photographs). The slopes (t-test) and residual variances (F-test) of these lines were compared to determine whether they differed significantly. Lack of difference among the lines indicated that the rate of growth has not changed significantly since 1946.

Each wetland on Gull Point is an individual case, and the authors felt that to attempt to calculate any statistics that grouped wetlands would seriously obscure the history of individual wetlands. Instead, the area of each wetland was plotted against year as a way to visualize temporal changes in wetland areas. To determine whether the area of an individual wetland changed significantly over time, least squares regressions were computed for each wetland, and the slopes of the regressions were computed to 0 using t-tests.

RESULTS

Variability of Measurements

For most wetlands, the standard error of the mean of independent measurements of surface area was small (Table 1), suggesting that single measurements obtained from the photographs were reliable estimates of the areas. For those wetlands with visually distinct edges (e.g., Wetlands 4 and 14), independent observers were in virtually complete agreement regarding the boundaries. For wetlands that were clearly defined but surrounded by dense stands of *Phragmites* (e.g., Wetlands 5, 12, 13, 16), the observers had no difficulty distinguishing the boundaries, but they did have problems outlining the wetlands accurately with the cursor. Most observers required several attempts at manually defining the outline, and the differences in measurements reflect, to a great extent, the skill/familiarity of the observer with the drawing apparatus.

Table 1 Means and Standard Errors for Areas of Seven Wetlands Present on the 1993 Photograph of Gull Point, Presque Isle State Park

Wetland	Mean area (ha)	Standard error	Maximum	Minimum
4	0.67	0.058	0.76	0.67
5	1.82	0.593	2.64	1.27
12	3.31	0.534	4.14	2.80
13	5.42	0.674	6.26	4.50
14	9.82	0.039	9.89	9.80
16	2.56	0.380	3.14	2.35
17	2.81	1.363	5.10	1.76

Note: Measurements were obtained from five different viewers with no previous experience with the software or photographs.

Wetland 17 presented serious problems for all observers. This wetland is surrounded by *Typha, Phragmites,* and mixed stands of *Populus* and *Salix.* It is difficult to find and penetrate the edges of the wetland even on foot, and the aerial photographs of the wetland were difficult to interpret. Most observers required a long period of consideration of the photograph before they were willing to attempt to draw in the borders manually. When they did draw the borders, they disagreed on what emergent vegetation was "in the wetland" and what vegetation constituted the upland edges of the wetland. Observers who actually had seen the wetland from the ground tended to give higher estimates of surface area than observers who had seen only the photograph.

All subsequent measurements of the wetlands were made by the senior author, who has conducted field work in all the wetlands under consideration and has had considerable practice using the image analysis system. This procedure was adopted to control among-observer variability.

Gull Point

In 1993, Gull Point was a 61.1-ha peninsula with approximately 3.8 ha of enclosed wetlands and an additional 3.9 ha of partially enclosed shallow lagoons (Figure 2, bottom). Of the 13 wetlands found on Gull Point in 1993, only one existed prior to 1974. Wetlands on Gull Point range from 0.06 to 1.0 ha in surface area and from several centimeters to approximately 1 m in depth. Wetlands less than three years of age have bare sand substrate with little or no surrounding vegetation. As the wetlands age, they become surrounded by dense stands of *Phragmites australis* up to 3 m in height that extend a short distance into the wetlands themselves, and 3 to 5 m onto the upland around the wetlands. *P. australis* rarely fills the wetlands, but it may cover shallow areas or ridges within them, and it appears to be an important agent in isolation of wetlands within the lagoon systems. Sand dunes and ridges between the wetlands are colonized by cottonwoods and upland vegetation, further isolating wetlands from one another.

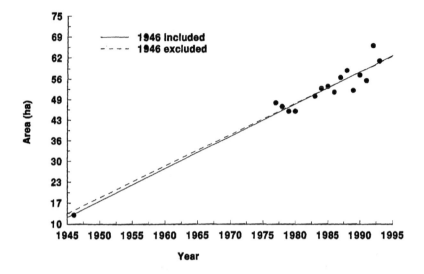

Figure 3   Plot of the least squares regressions of the total area of Gull Point, Presque Isle State Park, PA against the year. All areas were determined from aerial photographs of Gull Point taken at the same elevation and angle. Regressions were calculated with the 1946 data point included (solid line) and with the 1946 data point omitted (dashed line).

Areas

The area of Gull Point has increased significantly over the last 50 years (Figure 3), from 48.0 ha in 1946 to 61.1 ha in 1993 (Figure 2). Park records indicate that the rate of growth of Gull Point has become more rapid since 1955–1956 when aggressive beach renourishment began (Pennsylvania DER, 1993), but the data suggest that the rate of increase has been constant at least since 1946. The slope of a least-squares regression calculated with the 1946 area included does not differ significantly ($p > .05$, t-test) from the slope of the regression calculated with the 1946 area excluded (Figure 3).

The total area of wetlands on Gull Point ranged from a low of 2.0 ha in 1946 to a high of 6.7 ha in 1985 (Figure 4). Total wetland area increased between 1977 and 1985 and declined from 1985 to 1991. Since 1992, the total wetland area has been increasing. The total wetland area and lake level are significantly correlated ($p < .05$, $r^2 = 0.74$), but the areas of wetlands do not necessarily vary with the lake level on an individual basis (Figure 5).

As Gull Point wetlands aged, they did not necessarily decrease in area (Figure 5). Wetlands 4, 5, 6, 9, 13, 14, and 16 have shown no significant change in area over time, whereas Wetlands 10, 11, 12, and 17 have decreased in area. In those wetlands that decreased in area, the decline was not irreversible, and several had periods during which their areas increased. The increases and decreases were not correlated among wetlands ($p > .05$). That is, each wetland appeared to be on an individual trajectory in spite of hydrologic connections to each other and to the lake.

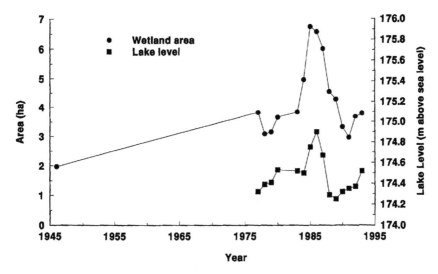

Figure 4   Plots of annual values for total wetland area on Gull Point, Presque Isle State Park, PA and Lake Erie annual mean elevation above sea level measured at the NOAA data station on Presque Isle.

Wetland Formation

Lagoons form on Gull Point when migrating offshore sandbars move toward shore at the end of the peninsula and become connected to the beach. Either the lagoons are destroyed when the bar is eroded by wave action or become closed at their bay ends, thereby forming wetlands. Eleven lagoon wetlands were identified on the 1993 photograph of Gull Point. The lagoon wetlands on Gull Point ranged in age from 6 to 47 years (Table 2), but 10 of them have formed since 1974. On average, new lagoon wetland systems, each consisting of 1 to 3 "sibling" wetlands, have been formed every 4 years (range 1 to 6 years). None of the lagoon wetlands formed since 1946 have disappeared, although Wetlands 10, 11, 12, and 17 have decreased significantly in area (Figure 5).

Beach wetlands form on Gull Point when incoming waves deposit sand on the beach, creating berms parallel to the shore. Water accumulates behind the berms and forms shallow ponds. If the ponds persist, they eventually become vegetated and form wetlands that closely resemble lagoon wetlands. The beach wetlands on Gull Point usually were ephemeral, appearing and disappearing within a single year. Since 1946, only four beach wetlands that had been formed persisted for more than 1 year. Wetlands 9.5 and 10.5 lasted for 4 and 3 years, respectively, before being destroyed by wave action (Table 2). Two beach wetlands, 8 and 8.5, formed in 1992, still are present on Gull Point.

DISCUSSION

Gull Point is a highly dynamic and rapidly evolving landform. It consists of a mosaic of wetlands ranging in age and, presumably, successional stage. Most of

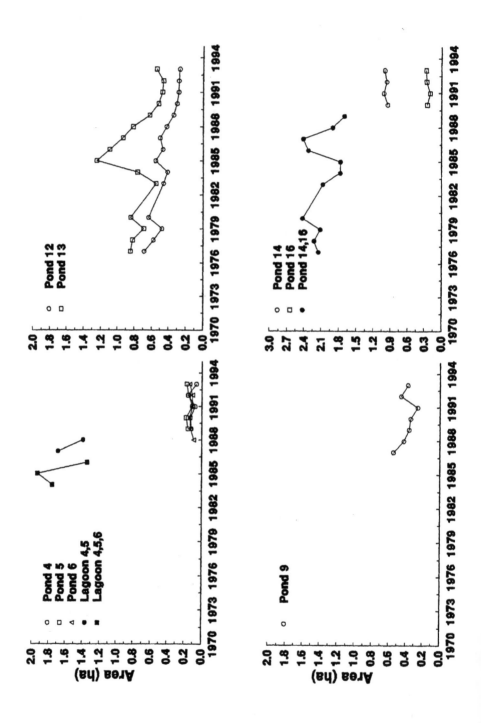

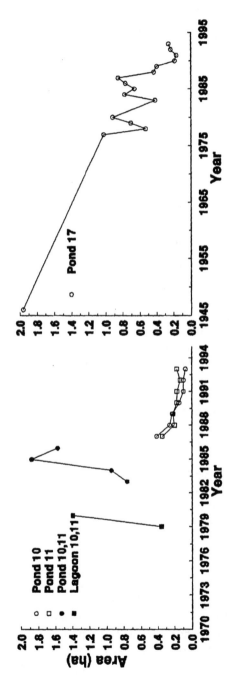

Figure 5 Plots of annual surface areas of wetlands on Gull Point, Presque Isle State Park, PA. Wetlands have been grouped on the basis of common origin. Thus, Wetlands 14 and 16 originally were united as a single Wetland 14,16. Solid symbols represent the united system, open symbols represent individual wetlands that formed from the united system.

Table 2   Year and Mode of Formation of Wetlands Found on Gull Point, Presque Isle, Pennsylvania From 1946 to 1993. Lagoons Are Identified by the Numbers of the Wetlands Formed From Them. Wetlands Are Numbered as in Figure 3

Wetland	Source	Formed	Isolated from lake	Separated from other wetlands	Destroyed
4	Lagoon 4,5,6	1984	1987	1989	NA
5	Lagoon 4,5,6	1984	1987	1989	NA
6	Lagoon 4,5,6	1984	1987	1987	NA
8	Beach	1992	1992	NA	NA
8.5	Beach	1992	1992	NA	NA
9	Lagoon 9	1986	1987	NA	NA
9.5	Beach	1983	1983	1983	1987
10	Lagoon 10,11	1979	1983	1987	NA
10.5	Beach	1984	1984	1984	1987
11	Lagoon 10,11	1979	1983	1987	NA
12	Lagoon 12,13	1975	1977	1977	NA
13	Lagoon 12,13	1975	1977	1977	NA
14	Lagoon 14,16	1974	1974	1989	NA
15	Lagoon 15	1948	NA	NA	NA
16	Lagoon 14,16	1974	1974	1989	NA
17	Lagoon 17	1946	1946	1946	NA
18	unknown	unknown	unknown	unknown	1980

it was not even in existence 20 years ago. Adequate characterization of the wetlands on Gull Point requires a technique that addresses both the current spatial boundaries of the wetlands and the temporal changes in those boundaries. Because the landform changes can be both rapid and dramatic, occurring within single seasons or storm events, the technique used to describe the history of the Gull Point wetlands should have a temporal resolution of one year or less. Additionally, to characterize the dynamics of all the wetlands on Gull Point, the technique used should provide a way to document the histories of wetlands that have persisted for long periods and allow for those wetlands that may have had prolonged periods of formation or that may have disappeared shortly after formation.

Wetlands on Presque Isle have been dated using a combination of tree ring analysis of cottonwoods growing on dune ridges between wetlands in conjunction with evaluation of the seral stage of the dune vegetation surrounding the wetlands (Jennings, 1909). To use this approach, one must assume that cottonwoods consistently colonize dunes between wetlands within two to three years of their formation (Jennings 1909; Ellenberger *et al.,* 1973) and that primary succession in and around the wetlands occurs within a predictable time scale and a predictable sequence of species establishment. Except for newly formed beach wetlands, the wetlands on Gull Point currently are surrounded by dense monocultures of *P. australis* which have overgrown other plant species including the cottonwoods (Botts, P. S., personal observation). Although cottonwoods and other plant species still occur on the dune ridges, the recent invasion of Gull Point by *P. australis,* an invasive form of which was introduced to Presque Isle in the 1970s, almost certainly has altered patterns of succession of vegetation on Gull Point from that described by Jennings (1909). As a consequence, the tree-ring/succession approach may be problematic in newer wetlands. Even in the absence of *P. australis*, the tree ring/succession approach has a temporal resolution of about 5 to 10 years at best. While tree ring analysis is useful in dating older wetlands on Presque Isle, it provides little information about wetlands on Gull Point, most of which are less than 20 years old.

Ellenberger *et al.* (1973) suggested that wetlands on Presque Isle could be dated using a biostratigraphic approach based on diatom assemblages in benthic core samples. However, to use a stratigraphic approach to date wetlands one must assume that diatom succession in wetlands is unidirectional, similar among wetlands, and begins at some discrete time of wetland formation. Our data suggest that wetland formation can be temporally prolonged, particularly when preceded by lagoon formation. Furthermore, wetlands frequently fluctuate in size and depth once they are established. Wetlands formed at the same time from the same lagoon show marked differences in organic content and heavy metals (Donn, R., unpublished data), as well as species composition and abundance of submerged aquatic macrophytes and invertebrates (Botts, P. S., unpublished data). It is unlikely that biostratigraphy can provide an accurate indication of the physical age of a wetland, particularly those of different depths, although it may well provide an indication of the "biological" age of a wetland.

The authors' initial interest in Gull Point was in its suitability as a site for answering questions on the influence of spatial and temporal patch dynamics on the species assemblages in the patches. A static picture of the spatial arrangement of the Gull Point wetlands as they are today provides little information about wetland history. It was necessary to clearly define the ages (temporal boundaries) and locations (spatial boundaries) of all wetlands on Gull Point in time as well as at the present day before ecologically relevant hypotheses could be formulated. Although research on Gull Point certainly would have proceeded in the absence of accurate dating, the types of questions asked now are much more specific and are less likely to be based on false premises.

For example, Lagoon 4,5,6 formed in 1984. Within this lagoon, Wetland 6 became isolated in 1987 and Wetlands 4 and 5 became isolated from each other in 1988. Wetland 6 presently is surrounded by bare sand and some *P. australis* and has a bare sand bottom; Wetland 5 is surrounded by dense *P. australis* and has a thick layer of organic muck above the sand substrate; Wetland 4 is surrounded by dense *P. australis* but still has a mixed sand and submergent vegetation substrate. The wetlands differ markedly in apparent successional stage and lack surrounding trees, making dating a difficult matter. In addition, the species composition and abundance of chironomid midges (Botts, P. S., unpublished data) in the three wetlands differs markedly. Without the aerial photograph data, the authors would have assumed a different time of origin (temporal boundary) for each wetland and they would have asked (inappropriately) how species composition and abundance of midges differed among wetlands of presumably different origin and age. With the photographic data, the authors were able to ask (appropriately) **why** physical characteristics such as sediment composition, and biological characteristics such as species composition and abundance of midges, differ among wetlands when the wetlands had a common origin.

Ecological research frequently is hampered by inadequate site histories. This can be especially troubling when the history itself is a datum of primary importance, as in studies of landscape ecology (Allen and Starr, 1982) or patch dynamics (Pickett and White, 1985). Although some of the necessary information can be inferred from indirect evidence, photographs provide unequivocal information regarding the history of a site. A series of photographs can provide information on both spatial and temporal boundaries. In addition, vegetation type and phenology or soil type frequently can be deduced from photographs (Lyon, 1993) and they provide a perspective not available from ground surveys.

ACKNOWLEDGMENTS

This study could not have been completed without the generosity of J. McKanna and R & M Biometrics, Inc. who kindly provided the image analysis software. H. Leslie of the Presque Isle State Park was most helpful in providing the aerial photographs and access to Gull Point. E. Masteller suggested that we use the photographs in the first place. This research was partially supported by an undergraduate research grant to R. Donn from The Pennsylvania State University — Erie.

REFERENCES

Allen, T. H. F. and Starr, T. B., 1982. *Hierarchy*. University of Chicago Press, Illinois.
Botts, P. S. and McCoy, E. D., 1994. Delineation of spatial boundaries in a wetland habitat. *Biodiversity and Conservation* 2:351–358.
Ellenberger, R. S., Laube, H. R., Walter, R. A., and Wholer, J. R., 1973. Pond succession on Presque Isle, Erie, Pennsylvania. *Proceedings of the Pennsylvania Academy of Science* 47:133–135.
Jennings, O. E., 1909. A botanical survey of Presque Isle, Erie County, Pennsylvania. Annals of the Carnegie Museum 5:145–159.
Kolasa, J. and Rollo, C. D., 1991. Introduction: the heterogeneity of heterogeneity: a glossary, in Kolasa, J. and Pickett, S. T. A., eds., *Ecological heterogeneity*, Springer Verlag, New York, pp. 1–23.
Kormandy, E. J., 1969. Comparative ecology of sandspit ponds, *American Midland Naturalist* 82:28–61.
Lyon, J. G., 1987. Maps, aerial photographs and remote sensor data for practical evaluation of hazardous waste site. *Photogrammetric Engineering and Remote Sensing* 53:515–519.
Lyon, J. G., 1993. *Wetland Identification and Delineation*, Lewis Publishers, Boca Raton, FL.
Lyon, J. G. and Drobney, R., 1984. Lake level effects as measured from aerial photographs. *Journal of Survey Engineering* 110:103–111.
Lyon, J. G. and Greene, R., 1992. Lake Erie water level effects as measured from aerial photographs. *Photogrammetric Engineering and Remote Sensing* 58:1355–1360.
Mitsch, W. J. and Gosselink, J. G., 1993. *Wetlands*. Van Nostrand Reinhold, New York.

National Oceanographic and Atmospheric Administration, 1994. Great Lakes water level data, monthly and annual average elevations. Washington, D.C.

Pennsylvania Department of Environmental Regulation, Office of Parks and Forestry, 1993. *Resource Management Plan: Presque Isle State Park*, Harrisburg, PA.

Pickett, S. T. A. and White, P. S., 1985. Patch dynamics: a synthesis, in Pickett, S. T. A. and White, P. S., eds., *The Ecology of Natural Disturbance and Patch Dynamics*, Academic Press, New York, pp. 371–384.

Reeder, B. C. and Eisner W. R., 1994. Holocene biogeochemical and pollen history of a Lake Erie, Ohio, coastal wetland. *Ohio Journal of Science* 94:87–93.

A COMPARISON OF WETLAND BOUNDARIES DELINEATED IN THE FIELD TO THOSE BOUNDARIES ON EXISTING STATE AND FEDERAL WETLAND MAPS
13   IN CENTRAL NEW YORK STATE

J. M. McMullen and P. A. Meacham

ABSTRACT

Wetland boundaries were delineated and surveyed in the field in a corridor approximately 47.5 km long and 152.4 m wide across Oswego County and the northern portion of Onondaga County in central New York State. These field-determined wetland areas were compared to hydric soils maps and mapped wetlands in the same corridor on federal and New York State maps. The maps were found not to agree and consistently underestimated the extent of wetlands; the federal National Wetlands Inventory maps were the least accurate, with the total wetland area underestimated by 61% compared to that delineated in the field.

INTRODUCTION

Existing wetland and soil survey maps produced by state and federal agencies are often used for environmental planning, development site selection, and linear facilities routing studies. The accuracy of these maps is frequently debated, with the producers of maps extolling their use and application and those involved with the field-delineation of wetlands expressing reservation. The intent of the current study was to compare field-delineated wetland boundaries to those on existing state and federal wetland maps and to maps of hydric (wetlands) soils. This comparison would provide an indication of the accuracy of these maps for the central New York State area.

1-56670-147-3/96/$0.00+$.50
© 1996 by CRC Press, Inc.

Detailed field studies of wetlands were performed by Terrestrial Environmental Specialists, Inc. (TES) in a linear corridor across Oswego County and northern Onondaga County in central New York State as part of environmental permitting for a cogeneration facility and a related overhead electric transmission line. Results of this study are presented in TES (1993). Comparisons were made between wetland areas delineated in the field and areas of hydric soil and wetlands shown on state and federal maps. This chapter provides the methods and results of this comparison. The comparison is based on total wetland area within the study area and highlights some specific differences in selected areas of the study area.

DESCRIPTION OF STUDY AREA

The study area extends across the towns of Scriba, Volney, Palermo, and Schroeppel in Oswego County and into the town of Clay in Onondaga County (Figure 1). The study area consists of a linear 61- to 183-m-wide corridor approximately 47.5 km long. For purposes of description, it is divided into segments (Figure 1).

Geology and Topography

All of Oswego County in the vicinity of the study area and the northern portion of Onondaga County is located within the Erie-Ontario Plain physiographic province (Stout, 1958). The regions within these two counties crossed by the study area are underlain by slightly tilted layers of Ordovician and Silurian sedimentary bedrock (SUNY, 1981; Miller, 1982).

The bedrock is overlain with thick glacial deposits. Glacial tills cover the northern portion of Oswego County, including most of the town of Scriba. Farther south, deposits are glacio-fluvial in origin. Glacio-lacustrine sediments occur in the southernmost portion of Oswego County and the northern portion of Onondaga County (USSCS, 1977; USSCS, 1981).

The topography of the area is largely the result of glacial deposition and erosion that followed glacial retreat. The northern portion of the county near the Lake Ontario shore is characterized by minor relief with slope profiles of only 0 to 3%. Beginning in the southern half of the Town of Scriba, elevations of the characteristically north-south oriented ridges reach 152.5 m and slopes may exceed 20%. In the southern portion of Oswego County and the northern portion of Onondaga County, the land is relatively flat with only subtle changes in elevation which range from 113 to 119 m asl.

Soils

A variety of soil types, each with distinct characteristics, have developed on the various glacial and glacio-lacustrine deposits in the study area (USSCS, 1977; USSCS, 1981). In the northern portion of the study area, Scriba gravelly fine

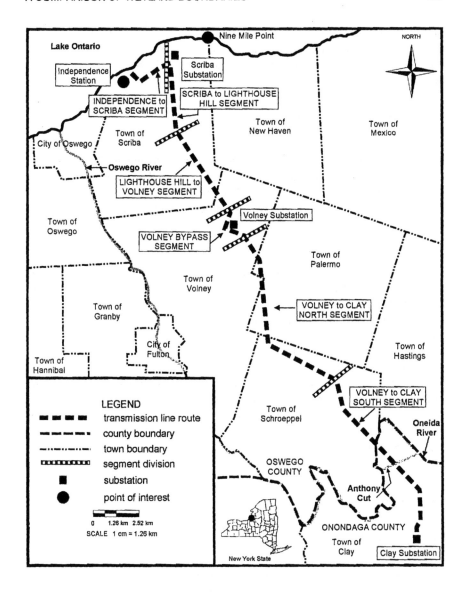

Figure 1 Location of study area.

sandy loam, Scriba very stony soils, and Ira-Sodus very stony soils are common in the uplands. Various types of hydric soils, which are predominantly Carlisle and Palms muck, occur in the central portion of the study area between Lighthouse Hill and Volney including the Volney Bypass area (Figure 1). Many of these same soils occur south of the Volney Substation, although there are large areas of sandy soils, primarily Hinckley gravelly loamy sand and Oakville loamy fine sand in the central portions of the Town of Schroeppel. The tight silt loams of Niagara and

Rhinebeck are common soil types in the study area in southern Oswego County and northern Onondaga County. These somewhat poorly drained soils occur on level ground with the hydric Canandaigua mucky silt loam present in slight depressions.

Surface Drainage Patterns

The study area crosses two major drainage basins. The Lake Ontario Basin includes all those streams which drain to the north directly into the lake. Drainage of the remainder of the study area is to the south through the Oswego River Basin, which includes the Oneida River and Anthony cut (Barge Canal). The Oneida River forms the border between Oswego and Onondaga County.

Surrounding Land Cover Types

For the most part the study area crosses rural open agricultural land in Oswego and Onondaga counties. Residential homes are scattered along the roads that cross the study area.

Land in the town of Scriba in the vicinity of the study area is primarily rural. Large areas of forest and patches of wetlands are mixed with low-density residential, agricultural, and industrial land (Bogner, 1979). In the Independence to Scriba segment, the major land cover is forest and wetlands.

Forest and brushland are the major land cover types along the Scriba to Volney segment, comprising over 85% of the existing corridor. Agriculture occupies about 8% of the land area with the remaining 7% in other categories (NMPC, 1982). Large muck fields occur in this part of the study area. The Volney to Clay segment contains large areas of active and abandoned agricultural land, forest, and wetlands.

Vegetation

The study area crosses a portion of the Eastern Ontario and Oneida Lake subregions of the Lake Plain region of New York (Stout, 1958). Common forest associations in this area are elm-ash-soft maple (*Ulmus-Fraxinus-Acer*) on poorly drained sites and beech-maple (*Fagus-Acer*) with a mixture of northern hardwoods on better drained upland sites (Stout, 1958). Eastern hemlock (*Tsuga canadensis*) is a common conifer tree species mixed with the deciduous trees or in dense evergreen stands.

Brushland or scrub-shrub communities are a major portion of the study area, partly because an existing cleared right of way constitutes about 75% of the study area and partly because abandoned agricultural land is common in the region. Abandoned agricultural land is particularly evident in the southern portion of Oswego County and northern portion of Onondaga County. Species of meadowsweet (*Spiraea* spp.), dogwood (*Cornus* spp.), brambles (*Rubus* spp.), and arrowwood (*Viburnum* spp.) are common dominants in these secondary communities and are usually mixed with open-field species such as common grasses and forbs.

Wetlands are common throughout the portion of Oswego and Onondaga counties crossed by the study area. Oswego County contains about 37,500 ha of wetlands regulated by the state (Jones et al., 1983). With a county land mass of approximately 250,000 ha, approximately 15% of the county contains state-regulated wetlands. With the addition of federally regulated wetlands, the percentage of wetland is likely to increase to 20 or 30%. According to USSCS (1981), Oswego County contains one of the largest areas of wetlands in the state with about 18,740 ha in muck land alone.

Most of the wetlands in the study area were originally forested. Since the study area contains a large portion of maintained rights-of-way, these areas of wetlands now contain emergent aquatic plants or shrub cover. Many of the shrub wetlands, particularly in the southern portion of the study area, have developed from abandoned agricultural land.

METHODS

Wetland boundaries were field-delineated in the study area beginning from near the shore of Lake Ontario in Oswego County and extending southerly across the entire county and into the northern portion of Onondaga County in central New York (Figure 1). The linear study area was divided into segments for purposes of the field delineation and to simplify data presentation. The location of the segments is shown in Figure 1. The study area often consisted largely of existing cleared rights-of-way. The width of the area ranged from 61 to 183 m and averaged 152.4 m. Details of the study area segments are as follows:

Segment	Areal extent
Independence Station to Scriba (IS)	61 m wide by 3.9 km long, 22.9 ha
Scriba to Lighthouse Hill (SL)	152.4 to 183 m wide by 5.8 km long, 90.4 ha
Lighthouse Hill to Volney (LV)	152.4 m wide by 7.6 km long, 113.7 ha
Volney Bypass (VB)	Approximately 91.5 m wide by 2.7 km long, 21.8 ha
Volney to Clay North (VCN)	152.4 m wide by 13.6 km long, 210.0 ha
Volney to Clay South (VCS)	152.4 m wide by 14.0 km long, 212.9 ha

Detailed data were collected and wetland boundaries were flagged within the segments of the study area by TES during June, July, August, and October of 1992. The boundaries were delineated using the federal criteria for vegetation, soils, and hydrology (Environmental Laboratory, 1987). Where New York State jurisdiction applied, state criteria were also taken into consideration. State-regulated wetlands were mapped primarily on the basis of vegetation.

Surveyor's ribbon was placed along the wetland boundaries based on the vegetation, soil, and hydrologic conditions. Boundaries, once flagged, were surveyed by C. T. Male Associates and plotted on 1:2,400 survey maps. Once the wetlands were accurately delineated, field surveyed, and the boundaries transferred to the maps, various types of information were determined for each wetland. The total area of each wetland within the corridor was determined using a CAD system.

The linear corridor used in the field study was accurately drawn on the soil survey maps of Oswego County (USSCS, 1981) and Onondaga County (USSCS, 1977), the New York State Department of Environmental Conservation (NYS DEC) freshwater wetlands maps of the two counties, and the National Wetlands Inventory (NWI) wetlands maps of the U.S. Fish and Wildlife Service. The soil surveys are based on field work conducted in the 1960s and 1970s, and used 1966 and 1975 aerial photographs as a base for the Oswego County and Onondaga County surveys, respectively. The wetland maps are produced on 7.5-minute U.S. Geological Survey (USGS) quadrangle maps at a scale of 1:24,000. The USGS quadrangles of West of Texas, Oswego East, New Haven, Pennellville, Central Square, and Brewerton were used. The NYS DEC wetland maps were filed with the Commissioner in 1986 and are based on wetlands mapping performed in the late 1970s and early 1980s. The NWI maps for these quadrangles are based on 1978 and 1980 aerial photographs at a scale of 1:80,000.

Hydric soils identified by the U.S. Soil Conservation Service (USSCS, 1989) were outlined in the study area, as were wetlands identified on the state and federal maps. The area of hydric soil and wetland was calculated by dot grid or by scaling directly from the maps. Total TES wetland areas were compared to total areas of mapped hydric soil, state wetlands, and federal wetlands.

A second more detailed comparison was made of TES field-delineated wetlands to those on state and federal wetlands maps in selected portions of the study area. Five plan drawings at a scale of 1:2,400 were selected at random for this comparison. Each plan covered about 1.6 km of the corridor and contained the wetlands flagged and surveyed in the field.

The selection of the plans was performed by study area segment to better represent the different characteristics of the linear study area. Segments IS, SL, and LV were combined, and one plan was selected from these segments. Two plans were chosen from segments VB and VCN. Two plans were also selected from the segment VCS, for a total of five selected plans.

The wetland boundaries from the NYS DEC wetland maps and the NWI maps were then transferred onto the selected plans. The total wetland area mapped by TES, the NYS DEC maps, and the NWI maps were measured using a combination of CAD and a dot grid system.

The differences between wetlands mapped by TES and those on the NYS DEC and NWI maps were then measured using the TES wetlands as the basis for comparison. Figure 2 is a schematic drawing of a wetland example showing the kind of analysis performed for this comparison. Areas mapped as wetlands outside of the TES wetlands were given a "positive" (+) designation and wetlands not mapped within the boundary of the TES wetlands survey were given a "negative" (–) designation.

The areas of each difference in wetland mapping, both positive and negative as related to the TES wetlands, were calculated. The total positive wetlands areas and negative wetland areas for both NYS DEC and NWI wetlands were then calculated for each plan.

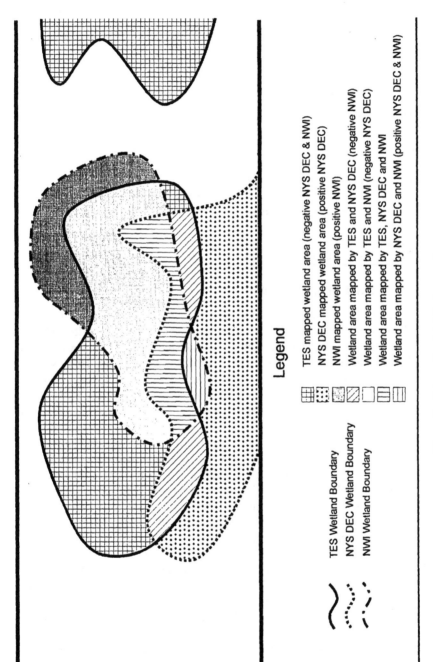

Legend

~ TES Wetland Boundary

⋯ NYS DEC Wetland Boundary

- - NWI Wetland Boundary

⊞ TES mapped wetland area (negative NYS DEC & NWI)

▦ NYS DEC mapped wetland area (positive NYS DEC)

▨ NWI mapped wetland area (positive NWI)

▧ Wetland area mapped by TES and NYS DEC (negative NWI)

▢ Wetland area mapped by TES and NWI (negative NYS DEC)

▤ Wetland area mapped by TES, NYS DEC and NWI

▥ Wetland area mapped by NYS DEC and NWI (positive NYS DEC & NWI)

Figure 2 Schematic of detailed comparison.

Table 1   Summary of TES Wetland Data by Study Area Segment

Area of wetlands (ha)	Area of cover types (ha)				Number of DEC wetlands	Number of streams	Number of flood plains
	OW	EW	SW	FW			
Independence Station to Scriba (IS)							
4.7	0.1		1.2	3.3	9	3	2
Scriba to Lighthouse Hill (SL)							
17.6	0.9	12.7	4.0		6	6	0
Lighthouse Hill to Volney Bypass (LV)							
28.1	1.3	16.6	5.7	4.5	8	7	1
Volney Bypass (VB)							
3.2		0.8	0.8	1.5	2	2	2
Volney Bypass to Clay : North (VCN)							
51.9	1.7	32.7	7.9	9.6	24	7	6
Volney Bypass to Clay: South (VCS)							
75.6	5.5	34.2	28.2	7.3	10	11	9
TOTALS:							
181.2	9.6	97.1	48.4	26.3	59	36	20

RESULTS AND DISCUSSION

A total of 233 wetland or open water areas representing a total of 181.2 ha were found by TES within the 47.5-km study area (Table 1). Although the delineation of wetland areas was confined to the study area limits, the wetlands usually extended beyond these limits. Each distinct wetland labeled by TES could have been part of a larger wetland, but was separated because of upland separation within the study area. The wetland areas were primarily emergent and scrub-shrub types, although this is largely due to the fact that a major portion of the study area is cleared right-of-way.

The distribution of wetlands by number and size was not uniform throughout the study area. Segments IS, SL, LV, and VB cover approximately one-third of the study area length and accounted for 60% (139 wetlands) of the total number of wetlands and only 30% (53.6 ha) of the wetland area. The last two segments, VCN and VCS, covered approximately two-thirds of the study area length, accounting for 40% (94 wetlands) of the total number of wetlands and 70% (127.6 ha) of the wetland area. The most common wetland cover types for the entire study area were emergent (EW) at 54% (97.1 ha), followed by scrub-shrub (SW) at 27% (48.4 ha), forested (FW) at 14% (26.3 ha), and open water (OW) at 5% (9.6 ha).

There are 59 wetlands identified by TES in the entire study area that are considered to be part of NYS DEC regulated wetlands. Twenty of the total 233 wetlands are within areas mapped as floodplains and 36 of them are associated with streams.

The total wetland area of 181.2 ha within the study area represents 27% of the 672-ha study area. With the assumption that this is a representative sample of the county, then about 27% of the county would be expected to be wetlands. This estimation of the total wetland extent in the county is comparable to the figure of 30% given in a report by NA (1991).

Table 2   Comparison of TES Field-Delineated Wetlands Data to
Mapped Hydric Soil and State and Federal Wetlands Data

| Study area segment characteristics | | TES wetland (ha) | Hydric soils (ha) | NYSDEC wetland (ha) | NWI wetland (ha) |
Segment	Length (km)	Area (ha)				
IS	3.9	22.9	4.7	2.2	4.4	4.0
SL	5.8	90.4	17.6	2.6	6.5	7.6
LV	7.6	113.7	28.1	18.9ᵃ	7.9	6.9
VB	2.7	21.8	3.2	6.1	5.0	2.9
VCN	13.8	210.0	51.9	52.8	46.1	28.4
VCS	14.0	212.9	75.6	25.1	44.2	20.7
Totals	47.8	671.6	181.2	107.7	114.2	70.5
Totals compared to TES wetlands		27%		(–)41%	(–)37%	(–)61%

ᵃ Excludes 20.4 ha of hydric soil currently being used for muck farm.

Comparison to Existing Maps

Results of the comparison of the total area of wetlands according to delineations by TES to those on existing maps is presented in Table 2. TES delineated a total of 181.2 ha of wetlands in the study area. Results of comparison for more specific areas in the study area are presented in Table 3.

Comparison of Total Wetland Areas

Mapped hydric soils total 107.7 ha in the study area, which is 41% less than the wetland area delineated by TES (Table 2). The mapped hydric soil area of 107.7 ha does not include about 20.2 ha of muck land used for agricultural purposes. These areas are effectively drained and none of the wetland maps recognizes these areas as wetland and they are not included as wetlands on the TES maps. Most of the errors in the soil survey maps are areas of wetlands that occur in soils mapped as having potential hydric inclusions, particularly the tight silt loams of Rhinebeck and Niagara soils in the abandoned agricultural land in the southern portion of the study area.

Mapped NYS DEC wetlands in the study area total 114.2 ha, which is 37% less than delineated in the field by TES (Table 2). Much of this difference is expected because the NYS DEC does not recognize wetlands smaller than 5 ha in size. As a result, area of wetlands mapped by TES can be expected to be greater than indicated on the NYS DEC maps.

Mapped NWI wetlands in the study area total 70.5 ha, which is 61% less than the total wetland area delineated in the field by TES (Table 2). The NWI maps, therefore, greatly underestimate wetland area for the study area. Although some smaller wetlands may not have been included on the NWI maps due to the large map scale, the considerable difference between the NWI maps to those field delineated by TES presents a major problem with NWI maps. The accuracy of the NWI maps for our study area in central New York State were considerably different than the "high degree of accuracy" reported by Wilen and Pywell (1992), or the 95% accuracy reported from Massachusetts by Swartwout et al. (1981). The

Table 3   Detailed Comparison on Selected Plans Between TES Field-Delineated Wetlands and Wetlands Mapped by NYS DEC and National Wetlands Inventory

Selected plan number	Total TES wetlands (ha)	Total NYS DEC wetlands (ha)	Total NWI wetlands (ha)	Total differences between TES and NYS DEC wetlands (ha)		Total differences between TES and NWI wetlands (ha)	
				Positive[a]	Negative[b]	Positive[a]	Negative[b]
5	6.8	2.5	4.2	(+)1.2	(−)3.5	(+)2.3	(−)3.5
10	14.2	12.2	4.6	(+)1.0	(−)5.5	(+)0.4	(−)5.8
13	5.4	3.7	1.2	(+)1.0	(−)3.3	(+)0.3	(−)4.6
17	2.9	3.3	2.9	(+)2.2	(−)1.1	(+)1.9	(−)2.2
22	16.1	12.2	1.2	(+)2.4	(−)6.8	(+)0.2	(−)7.1
Totals	45.5	33.9	14.2	(+)8.2	(−)20.2	(+)5.2	(−)23.2

[a] Areas where wetlands are larger than those mapped by TES.
[b] Areas where wetlands are smaller than those mapped by TES.

deficiencies in the NWI maps are fairly uniform throughout the study area. In the southern portion of the study area several large wetlands were not indicated on the NWI maps, although they were included on the NYS DEC maps.

Detailed Comparison of Wetlands on Selected Plans

A comparison of the total area of wetlands can be misleading because it does not reflect where the area of mapped wetlands differs from the TES field-delineated wetlands by being larger in one area and smaller in another and thus balances the total amount. To address this problem, five plans were selected at random to compare specific positive and negative differences in more detail.

Results of the detailed comparison on the five selected plans (about 8 km of the 47.5-km study area) are presented in Table 3. Total TES wetlands on these plans was 45.5 ha, with a total NYS DEC wetlands of 34.0 ha and a total NWI wetlands of 14.2 ha.

The NYS DEC wetlands on these plans totaled 34.0 ha compared to 45.5 ha determined by TES (Table 3). Total area where the NYS DEC mapped wetlands in addition to TES (positive differences) was +8.2 ha and the total areas where the NYS DEC wetlands were less than TES (negative differences) were –20.2 ha. As a result, the NYS DEC maps not only considerably underestimated the amount of wetlands, they were also often found to be in error. Although the NYS DEC regulate and map wetlands greater than 5 ha in size, this was considered in our analysis and so negative differences between the NYS DEC maps and TES maps were expected.

In the detailed comparison of the five plans, the NWI maps were found to be considerably in error and generally underestimated the wetlands area. The total area of NWI wetlands was only 14.2 ha compared to the 45.5 ha mapped by TES. Specific positive differences were (+)5.2 ha and negative differences were (–)23.2 ha. As found with the comparison of total wetlands in the study area, the NWI maps significantly underestimated the amount of wetlands.

CONCLUSIONS

Within a broad belt about 47.5 km in length in central New York State, this study found that existing soil survey maps and state and federal wetland maps severely underestimated total wetland extent when compared to more detailed surveys in the field. Compared to the TES wetland maps delineated in the field, the hydric soils, NYS DEC, and NWI maps showed 41, 37, and 61% less total wetland area, respectively. The existing wetlands maps were also found to be inaccurate, indicating more wetland extent in some areas and less in others, with an overall estimation of less wetland in the study area.

There may be several explanations for inaccuracies in existing maps. The major reasons are probably the different scales used in mapping wetlands and the scale and quality of aerial photographs used to interpret and delimit wetlands.

Jurisdictional limitation was a major limitation in the case of the NYS DEC maps, since this agency is only concerned with wetlands greater than 5 ha in size. Surprisingly, of the wetlands included on the NYS DEC maps, fewer discrepancies were noted when compared to the TES field-delineated wetland maps.

The results of this study not only draw attention to the weaknesses and inaccuracies of existing wetland maps, they also show the special care necessary when using existing maps in wetland trend analysis. Tiner (1985) and Moorhead and Cook (1992) compared areas of hydric soil survey maps to more recent NWI maps. They found less wetland area on the NWI maps than the area of hydric soil and attributed the smaller estimate based on NWI maps as evidence for a net reduction in wetland extent. The results of the present study point out that the existing wetland maps are not reliable enough to draw conclusions about losses or gains in wetland extent.

ACKNOWLEDGMENTS

The field delineation and compilation of wetlands maps were part of permitting studies supported by Sithe/Independence Power Partners, L. P. (Sithe), and Niagara Mohawk Power Corporation (NMPC). Their contributions to this study are acknowledged.

REFERENCES

Bogner, J., 1979. A Natural Resource Inventory for the Town of Scriba, NY. Oswego County Planning Board, Environmental Management Council and Massachusetts Audubon Society.

Environmental Laboratory, 1987. Corps of Engineers Wetlands Delineation Manual. *Technical Report Y* 87 1, U.S. Army Engineer Waterways Experiment Station, Vicksburg, MS.

Jones, S. A., Corey, M. E., and Zicari, L., 1983. The Oswego County Wetlands Mapping Inventory Project: Introduction and Summary. Oswego County Environmental Management Council, Oswego, New York.

Miller, T. S., 1982. Geology and Ground Water Resources of Oswego County, New York. U.S. Geological Survey Water Resources Investigation 81 60, in Cooperation With Oswego County Planning Board.

Moorhead, K. K. and Cook, A. E., 1992. A comparison of hydric soils, wetlands and land use in coastal North Carolina. *Wetlands* 12:99–105.

NA, 1991. Wetlands Status and Trends Analysis in Oswego County, New York. Prepared for U.S. Environmental Protection Agency by Normandeau Associates, Inc., Toms River, NJ.

Niagara Mohawk Power Company, 1982. Nine Mile 2 Volney 345 kV Transmission Facility. Niagara Mohawk Power Corporation, Syracuse, NY.

Stout, N. S., 1958. Atlas of Forestry in New York. State University College of Forestry at Syracuse, Bull. No. 41, Syracuse, NY.

SUNY, 1981. Geology of New York. Education Leaflet No. 20, New York State Museum and Science Service, Albany, NY.

Swartwout, D. J., MacConnell, W. P., and Finn, J. T., 1981. An Evaluation of the National Wetlands Inventory in Massachusetts. In: *Proc*. In: *Place Resource Inventories Workshop*, 9–14 August 1981, University of Maine, Orono, ME, pp. 685–691.

TES, 1993. Wetlands Delineation Report, Independence Station Clay 345 kV Transmission Line Project. Terrestrial Environmental Specialists, Inc., Phoenix, NY.

Tiner, R. W., Jr., 1985. *Wetlands of New Jersey*. U.S. Fish and Wildlife Service, National Wetlands Inventory, Newton Corner, MA.

USSCS, 1977. Soil Survey of Onondaga County, New York. U.S.D.A. Soil Conservation Service in Cooperation with Cornell University Agricultural Experiment Station.

USSCS, 1981. Soil Survey of Oswego County, New York. U.S.D.A. Soil Conservation Service in Cooperation with Cornell University Agricultural Experiment Station.

USSCS, 1989. Hydric Soils of the State of New York. U.S.D.A. Soil Conservation Service in Cooperation with National Technical Committee for Hydric Soils, Washington, D.C.

Wilen, B. O. and Pywell R. H., 1992. Remote Sensing of the Nation's Wetlands, National Wetlands Inventory. In: *Proceedings; Fourth Biennial Forest Service Remote Sensing Applications Conference,* 6–10 April 1992, Orlando, Florida.

14 WETLAND BUFFERS AND RUNOFF HYDROLOGY

J. D. Phillips

ABSTRACT

Wetlands are hydrologic buffers by virtue of their locations within landscapes and may serve a variety of buffer roles, including that of a water quality filter strip. Water quality buffer effectiveness (with respect to storm runoff from adjacent land) depends on the ability or propensity to (1) delay flow or reduce flow velocities through the buffer; (2) reduce or minimize the stream power of overland flow; (3) produce surface runoff; and (4) maintain particular biogeochemical conditions which are pollutant specific. In addition, for riparian buffers the relative proportion of water supplied from runoff from adjacent hillslopes vs. overbank flooding is a critical consideration. In general, wetlands are inferior to nonwetlands with respect to delaying flow and producing surface runoff. Wetlands are often superior for reducing stream power and may be more or less effective than nonwetlands with respect to specific biogeochemical conditions, depending on whether aerobic or anaerobic processes are required. This is demonstrated by comparing buffer effectiveness indices of hydric and nonhydric soils for 161 soil series of the Tar River basin, North Carolina. The water quality buffer values of wetlands derive primarily from their landscape setting and vegetated status, not from their hydrologic properties. Because wetlands may be poor buffers and are themselves often critical resources, wetlands themselves should be buffered in many cases.

INTRODUCTION

Wetlands, by their very nature, are hydrologic buffers. They buffer the effects of upland runoff on ground and surface water bodies and the effects of high water, floods, and tides on uplands. Wetlands invariably function as filters and valves in

1-56670-147-3/96/$0.00+$.50

the hydrologic system, because they cannot form except in land/water transition zones, or (in humid climates) in areas where there is a net moisture surplus.

These buffering roles of wetlands are well known in hydrology. However, given the widespread efforts to protect wetlands, numerous threats to wetland integrity, and the increasingly apparent impacts of human agency on the hydrologic system, new questions arise about the hydrologic buffering functions of wetlands. How can these buffer functions be used, actively or passively, to protect and enhance water resources? How effective are wetlands as buffers, as compared to other, nonwetland environments? To what extent should wetlands themselves be buffered? The purpose of this chapter is to explore these questions. The focus will progressively narrow, from the purposes of buffers, to the factors determining riparian water quality buffer effectiveness, to the relative hydrologic effectiveness of wetlands as buffers for adjacent land. This focus ignores other important buffering roles of wetlands, but does address questions which are currently critical to resource managers (Budd *et al.*, 1987; Diamond and Nilson, 1988; Dodd *et al.*, 1993; Nieswand *et al.*, 1990; Phillips, 1989c,d).

Why Riparian Buffers?

Establishing and maintaining vegetated riparian buffer zones, wetland or otherwise, generally has one or more of four general purposes. The first is water quality protection, for which there are two main aspects — displacement and filtering. Riparian buffers spatially displace potential runoff and pollution sources from the water body. Because the travel time of runoff or effluent varies as the square of the length of flow (Phillips, 1989a), displacement allows for attenuation of many contaminants. Riparian buffers serve as filters due to the capacity of soil, vegetation, and microorganisms to remove and/or transform pollutants. The second major purpose of vegetated riparian buffers is to provide habitat — provision of undeveloped corridors for wildlife movement and maintenance of specific riparian habitats. A third major purpose is for erosion setbacks. This is common in oceanfront settings, but North Carolina's 75-ft estuarine shoreline "area of environmental concern", for example, was also enacted as an erosion setback (Phillips, 1989c). Finally, riparian buffer zones are sometimes established to provide shoreline or streambank access or recreation space, or for the mundane purpose of maintaining property line setbacks (Phillips and Phillips, 1988).

Some generic types of programs requiring or encouraging riparian buffers or regulating riparian land use are listed in Table 1. The remainder of this chapter will focus on the water quality role of wetland buffers.

WETLANDS AS WATER QUALITY BUFFERS

The values of riparian wetlands as water quality buffers are a function of the four criteria below, any of which may be of paramount or minimal importance for specific objectives or locations.

Table 1   Generic Types of Riparian Buffer Regulations and Programs

Buffer Program	Type[a]	Purpose
Minimum vegetated buffer adjacent to critical waters (high quality, critical habitats, protected waters, water supply, etc.)	S,I P	Water quality Habitat
USDA Conservation Reserve Program, streamside zones	I	Water quality Habitat
Estuarine shoreline management zones (for example, Maryland Critical areas program; state coastal management programs)	P	Erosion setback Water Quality Habitat Access
Onsite wastewater treatment regulations; minimum setbacks from surface waters	S	Water quality
Land application of sludge, wastewater, animal waste, etc. regulations; minimum setbacks from surface waters	S	Water quality
Nonpoint source pollution "best management practice" cost share programs including filter strips	I	Water quality
Local subdivision regulations; property line minimum setbacks	S	Access

[a] I = incentives, such as cost-sharing or cross-compliance; S = mandated minimum setbacks; P = permits required for development.

Ability to delay flow — As a rule, the longer it takes runoff or effluent to pass through the buffer, the more effective the buffer is. The velocity of a given volume of overland flow entering the buffer will depend on infiltration rates within the buffer; surface roughness derived from microtopography, coarse clasts or aggregates, vegetation and organic litter; and slope gradient. Rapid hydraulic conductivity promotes infiltration and tends to slow overall and surface flow rates, but enhances subsurface flow rates, which depend on conductivity and water table slopes. Wetlands, as compared to nonwetland buffers, generally have high hydraulic roughness and low slopes, but may allow minimal infiltration due to saturated or near-saturated soils.

Ability to reduce or minimize stream power — Stream power defines the bulk solid mass transport capacity of surface flow through the buffer and is particularly relevant to sediment and other particulates. Stream power is reduced by higher infiltration rates, shorter slope lengths, greater hydraulic roughness and, especially, gentler slope gradients. Because slope gradients have greater relative importance than infiltration with respect to stream power, wetland buffers generally perform well with respect to this criterion.

Propensity to produce runoff — Within a particular climatic regime, the amount of runoff produced from a given precipitation event is primarily a function of infiltration and depression storage. The latter is substantial in some wetlands, but infiltration rates are often low due to soil saturation, water tables at, near, or above the surface, and low hydraulic conductivity of some wetland soils. Many riparian wetlands are critical runoff *source* areas (Troendle, 1985). Wetlands fare poorly relative to nonwetlands on this criterion.

Ability to remove or treat specific pollutants — Some pollutants, such as heavy metals, some pesticides, and phosphorus are transported primarily in association with fine-grained sediments. Thus the ability of the buffer to remove these pollutants is largely controlled by stream power. Other pollutants, such as bacterial pathogens and oxygen-demanding wastes, can be treated effectively by delaying flow, thus allowing time for die off and decomposition. In other situations, specific biogeochemical processes such as denitrification are critical and require specific chemical as well as hydrological conditions. Where processes such as denitrification require anaerobic conditions common in wetlands, the latter are superior buffers. Where aerobic conditions are required, wetlands are inferior buffers.

Even where factors such as low slopes and anaerobic conditions are most critical to wetland buffer functions, there are serious questions as to the values of wetlands as stormwater buffers. There is little question that many wetlands serve as long-term storage/sink areas for various pollutants and that they are effective with respect to ground water flow. However, wetland buffers may be minimally effective during flood or runoff peaks when the largest "flushes" of contaminants are typically delivered. Nitrates cannot be denitrified, for example, if the stormwater is conveyed rapidly across or through the wetland. Devito and Dillon (1993), for example, have documented low nutrient assimilation during wet periods in a conifer swamp. Magette and others (1989) show that nutrient reductions in vegetated filter strips are highly variable and decline as the number of runoff events increases. This is a fruitful area of research for isotope hydrologists, who may be able to shed light on the extent to which "old" denitrified water from riparian wetlands is displaced by nitrate-laden stormwater which may then be treated.

A fifth criterion is linked to the landscape setting rather than to properties of the wetland itself: the relative importance of local vs. upstream runoff for riverine wetlands, or of upland runoff vs. flooding for lacustrine or estuarine shoreline wetlands (Brinson, 1993). This is now discussed.

LOCAL AND UPSTREAM RUNOFF SOURCES

With regard to shoreline wetlands, it is probably safe to assume that the amount of flooding from the water body is directly proportional to the hydroperiod and that filtering of upland runoff is of primary importance. With riverine wetlands, the extent to which the major concern is downstream attenuation of floodwater pollutants or filtering of runoff from adjacent uplands may determine the value of a particular buffer. In general, the lower the stream order (i.e., the closer one is to headwaters and drainage divides) and the smaller the upstream drainage area, the higher the ratio of local to upstream water sources. In first-order tributaries the adjacent hillslopes comprise virtually 100% of the water source, though except in very poorly drained terrain these basins are unlikely to support wetlands. This derives from the basic nature of fluvial drainage networks (Richards, 1982).

For a threshold event which produces both overbank flooding and runoff on adjacent uplands, an index of the relative importance of local and upstream sources (R) is given by

$$R = (q_h A_h)/(q_b A_b - Q_{bf})$$
(1)

where q is the mean runoff rate per unit area, A is the drainage area, and Q_{bf} is the bankfull discharge. The subscript h refers to the local hillslope draining to the wetland and b to the upstream drainage basin. Where there is overbank flow on the opposite side of the channel this should be added to Q_{bf}. While data- or process-based model estimates of q may be difficult and expensive to obtain, rough estimates of q can be based on simple methods such as the rational method or runoff curve numbers as described in hydrology textbooks (for example, Shaw, 1988; McCuen, 1982). Drainage area can be measured from topographic maps (or calculated from digital elevation models using standard algorithms). Bankfull discharge can be estimated from stream gage data or calculated from simple site measurements using the Manning or D'Arcy-Weisbach equation.

Events which produce hillslope runoff but not overbank flooding are irrelevant with respect to Equation 1 because all the water is then derived from local runoff. Likewise, conditions which produce sustained flood peaks well beyond the period of local runoff generation, or where storms and runoff are confined to upstream areas, are irrelevant because all water is derived from overbank flooding. A single index for any time period could be derived from

$$R = \left[\sum_{}^{n} (q_{hi} t_i A_h) \right] / \left[\sum_{}^{m} (q_{bj} t_j A_b) \right]$$
(2)

where there are i = 1,2,3,...n events which produce runoff on the hillslope, each with duration t_i, and j = 1,2,3,...m overbank flood events, each with duration t_j.

Obtaining reliable information for Equation 2 would require extensive data collection over a period sufficient for calculation of means or probability distributions, or extensive modeling. A simpler approach for obtaining a relative (rather than an absolute) index R' is to identify a design recurrence interval r (for example, two years). Then

$$R' = (q_{h,r} t_{h,r} A_h)/(q_{b,r} t_{b,r} Ab - Q_{bf} t_{b,r})$$
(3)

WETLAND AND NONWETLAND BUFFERS

Any riparian area which is flooded often enough to provide for buffering of floodwaters is likely to be a wetland — if not in some particular legal or regulatory definition, then in most hydrologic or geomorphic definitions (cf. Phillips, 1989e). The question of the extent to which wetlands are effective buffers as compared to nonwetlands is thus relevant only to the problem of buffering runoff from adjacent land.

The general effectiveness of wetland and nonwetland buffers can be evaluated by comparing wetland and nonwetland soils. Some controls of buffer effectiveness are directly related to soil properties. These include hydraulic conductivity, moisture storage capacity, and transmissivity. Others, such as infiltration capacity, are strongly controlled by soil properties (i.e., hydraulic conductivity) and, in a general sense, vary proportionally to soil properties. Still others (slope), while not direct soil properties, are systematically linked to particular soil types and are generally included in soil information databases. If one then assumes identical vegetation cover with respect to surface roughness and holds buffer width constant, the relative buffer effectiveness of soils can be calculated (see Phillips 1989b;c).

For the purposes of this study, all recognized soil series mapped in the 21,200-km^2 Tar River basin of North Carolina (Figure 1) are identified. The Tar River rises in the NC Piedmont province and flows across the Coastal Plain to the Pamlico estuary. The environmental setting of the basin is described elsewhere (PTRF, 1991; Copeland and Riggs, 1984). As the entire basin is not yet mapped, additional soils known to occur in similar settings in the state (Daniels *et al.*, 1984) were added. For each of the 161 soil series thus identified, soil information was obtained from the U.S. Soil Conservation Service's National Soils Database. This database, commonly referred to as SOILS-5, is described in SCS (1985). Soil series were divided into wetland and nonwetland categories based on their hydric status (SCS, 1987).

Each series was evaluated using an index reflecting the four buffer effectiveness criteria above. The ability to detain flow was assessed based on the detention time version of the Riparian Buffer Delineation Equation (RBDE). Development and calculation techniques are described elsewhere (Phillips, 1989a,b):

$$B_b/B_r = (L_b/L_r)^2(K_b/K_r)^{0.4}(n_b/n_r)^{0.6}(S_b/S_r)^{-0.7}(C_b/C_r) \qquad (4)$$

The term at the left is a buffer effectiveness ratio relating the site or soil to be evaluated (b) with some reference buffer (r). L is slope length or buffer width, K is saturated hydraulic conductivity or permeability, n is the Manning roughness coefficient, S is slope gradient, and C is soil moisture storage capacity. The equation is based on standard hydrologic analysis methods, using the Manning and D'Arcy-Weisbach equations. The major assumptions not necessarily inherent in those equations are that the ratio of the water table and surface slopes are identical, and that infiltration capacity is proportional to hydraulic conductivity.

In this situation L and n are assumed constant. The reference condition is based on properties of the Norfolk series, the single most common upland soil in the NC Coastal Plain. The Norfolk series is a fine loamy, siliceous thermic Typic Kandiudult which occurs in well-drained sites. SOILS-5 data give permeability or hydraulic conductivity, available moisture capacity, and slope gradients in ranges.

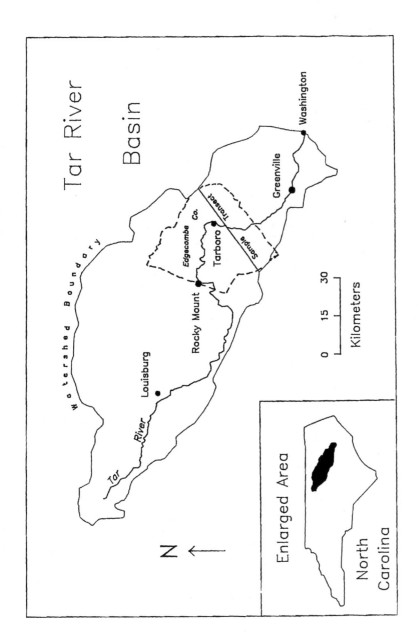

Figure 1 Map showing the location of the Tar River basin, North Carolina, U.S.

The median of the range was used, as previous studies have shown that the relative buffer effectiveness of different soils or sites is similar, regardless of whether maximum, minimum, or mid-range values are used (Phillips, 1989b,c,d). C is calculated by multiplying available water capacities for those layers of the representative profile above bedrock or a seasonal high-water table by the layer thickness and summing.

The ability of the buffer to reduce stream power was assessed on the basis of the hydraulic version of the RBDE. This is described in Phillips (1989a) and is based on the work of Moore and Burch (1986) in modeling unit stream power of overland flow.

$$B_b/B_r = (L_b/L_r)^{0.4}(K_b/K_r)(n_b/n_r)^{0.6}(S_b/S_r)^{-1.3} \qquad (5)$$

The Norfolk series was again used as the reference condition, with L and n held constant.

Propensity to produce runoff was assessed using the wetness index (WI) presented by Phillips (1990). The latter is one of several such indices which integrate the influences of upslope drainage area, flow convergence or divergence, and soil transmissivity. While the index below is derived from the hydrologic model of Beven and Kirkby (1979), other, similar indices are available (for example, O'Loughlin, 1986). The approach here is based on the following expression for moisture storage deficit s (Beven, 1986):

$$s_i = s + m \lambda - m \ln(aT_e/T_i \tan B) \qquad (6)$$

The moisture storage deficit at a given point in a drainage unit (such as a hillslope or watershed), s_i, is a function of the mean deficit for the whole unit (s), the moisture deficit with regard to the saturated value (m), the area drained per unit contour length at point i (a), the local and mean soil transmissivities (T_i and T_e), and slope angle (B, in degrees); λ is a parameter related to the integral of transmissivity ratios over the drainage area. If $s_i < 0$ there is no storage deficit, i is saturated and any precipitation or runon becomes saturation-excess runoff. Full derivations are given elsewhere (Beven and Kirkby, 1979; Beven, 1986), and a fuller treatment in the context of wetness indices is given by Phillips (1990).

The latter term of Equation 6, $\ln(aT_e/T_i \tan B)$, is an index of hydrologic similarity with regard to saturation. The relative wetness of any two points within a drainage unit can be compared on the basis of this index. If one is concerned only with relative wetness, T_e can be arbitrarily set to unity. Then, after simplification, a simple wetness index is derived (Phillips, 1990):

$$WI = \ln[a/(T_i \tan B_i)] \qquad (7)$$

This index expresses wetness as a function of upslope drainage area, transmissivity (itself a function of hydraulic conductivity and depth to a water table or confining layer), and slope gradient. Because soils data were utilized, each i is a soil series and

$$T_i = \sum^{D_{max}} K_z D_z \tag{8}$$

where K is the permeability or saturated hydraulic conductivity of each horizon z or portion thereof above a confining layer or seasonal high-water table, D_z is the thickness of the horizon, and D_{max} is the depth from the surface to the confining layer or water table. Depth of the seasonal high-water table is indicated in the SOILS-5 data and is also usually reflected in the soil profile morphology of the typical pedon description. Where water tables are at or above the surface, a minimum D_{max} of 5 cm (2 in.) was used for calculations. Because WI requires site-specific topographic information the SOILS-5 data could not be used. Rather, data from a transect across the Tar River basin including 25 soil types at 324 sample sites from drainage divide to drainage divide were utilized. The transect and data collection are described elsewhere (Phillips, 1990).

Higher values of the wetness index indicate wetter sites which are more likely to produce saturation excess runoff. The propensity to produce infiltration-excess runoff can be assessed by comparing transmissivity. The latter is calculated by summing the product of permeability and layer thickness for each layer above bedrock or a seasonal high-water table (see Equation 8). A limiting depth of 1.83 m (6 ft) was imposed for calculations. The ratio of the Norfolk transmissivity to that of each soil was used, so that low values indicate soils which transmit moisture most effectively and are thus less prone to produce surface runoff.

The ability to filter specific pollutants tends to be a function of flow detention, stream power of overland flow, or maintenance of aerobic (unsaturated) or anaerobic (saturated) conditions. Therefore, the detention and hydraulic versions of the RBDE and the wetness index can be used for this purpose.

RESULTS

Of the 161 Tar River Basin soil types analyzed, 98 are nonhydric and 63 are hydric. Means and standard deviations for the indices are shown in Table 2. Tests for significant differences were conducted using Z-scores (Table 3). Means for hydric and nonhydric soils were compared to the overall mean for all 161 soils. Nonhydric and hydric means were compared to each other with two separate tests.

Buffer Effectiveness Ratio

The buffer effectiveness ratio for nonhydric soils was both higher and more variable than for hydric soils. The means for the two groups are different from each other at >99% confidence and from the overall mean of 0.443 at the same confidence level. The majority of hydric soils had ratios less than 0.3 and none were greater than unity.

Table 2   Indicators of Buffer Effectiveness
(Overall Mean Value is in Parentheses)

Variable	Hydric soils		Nonhydric soils	
	Mean	Std. dev.	Mean	Std. dev.
Buffer Effectiveness Ratio (0.443)	0.176	0.087	0.610	0.454
Wetness Index[a] (–4.708)	–1.590	2.240	–4.950	1.810
Transmissivity Ratio (17.710)	41.669	92.192	2.305	3.412
Sediment Buffer Index (1.695)	2.831	3.254	1.003	1.741

[a] Based on a sample transect across the Tar River basin involving 324 sample points and including 25 soil types. See text.

Table 3   Tests for Significant Differences

Variable	H:NH	NH:H	H:X	NH:X
Buffer Effectiveness Ratio	–39.60	9.46	–24.36	3.64
Wetness Index	3.71	3.71		
Transmissivity Ratio	3.39	–114.21	2.06*	–44.69
Sediment Buffer Index	5.56	–8.33	3.46	–3.15

Note: All are based on z-scores except for the wetness index, which is based on the Student's t-test. Comparisons are hydric vs. nonhydric and vice versa (H:NH and NH:H) and hydric and nonhydric means vs. the overall mean (H:X and NH:X). All differences are statistically significant at a probability of >99% except the value marked *, which is significant at the 95% confidence level.

Wetness Index

The transect included 324 sample points, at which 25 different soil types were found (17 nonhydric; 8 hydric; see Phillips, 1990). Mean wetness index values for those soils were calculated and the differences between the hydric and nonhydric soils tested (using the Student's t-test in this case due to the smaller sample size). The hydric soils clearly show the greatest potential for producing saturation-excess runoff and for maintaining anaerobic conditions, though WI was considerably more variable than for the nonhydric soils. The nonhydric mean of nearly –5.0 was statistically different from the hydric mean of 1.6 at 99% confidence. Mean values for all nonhydric soil sites were all less than the overall mean of the 25 soils of 4.71, while means for the hydric sites ranged from 1.26 to –4.75.

Transmissivity Ratio

The transmissivity ratio averaged almost 18 for the entire data set. High values are attributed to the fact that the Norfolk has a deep water table (>1.8 m), coupled with high permeability in the surface horizons. Most other coastal plain soils have higher water tables, while piedmont soils are typically thinner and may have lower permeability. The mean of almost 42 for hydric soils was statistically different from the mean with 95% confidence; the nonhydric mean of about 2.3

differed from the overall mean with >99% confidence. The hydric and nonhydric means were statistically different at >99% confidence.

Sediment Buffer Index

Hydric soils are generally superior buffers of sediment-transporting overland flow, with a mean index about 2.8 times higher than that of the nonhydric soils. Though the standard deviations are not excessive (see Table 2), both means are influenced by a few very large (>5) values. Both the hydric and nonhydric means are different from the overall mean and from each other at >99% confidence.

To sum up, hydric and nonhdyric groups of soils from the Tar River Basin samples clearly fall into two distinct populations with respect to hydrologic properties. Hydric soils are more likely than nonhydric soils to produce both saturation-excess runoff (as indicated by the wetness index) and infiltration-excess overland flow (as indicated by the transmissivity ratio). While hydric soils are superior buffers with respect to reducing stream power and thus for sediment and other coarse particulates, the nonhydric soils are generally preferred in terms of delaying flow.

DISCUSSION

Except in special situations where particular goals such as maintenance of anaerobic conditions or reduction of overland flow stream power are sought, wetlands and hydric soils are generally less effective buffers than upland, nonhydric soils. The water quality buffer advantages of wetlands do not, therefore, arise from any properties associated with wetland hydrologic properties. Rather, they arise due to the strategic locations of wetlands between runoff and effluent sources and waterways, and the fact that wetlands are far less likely to be developed than uplands. In other words, the hydrologic values of wetland buffers derive primarily from their landscape setting and not from wetland hydrology.

This has several implications for wetland and riparian zone management, presented here as recommendations. First, vegetated riparian areas should be preserved or managed as water quality buffers. This management or preservation should not be contingent upon status as a jurisdictional or any other type of designated wetland. As a corollary, riparian wetlands should not (from a runoff buffer perspective) be accorded any priority over nonwetlands in programs to require, acquire, encourage, or manage buffer zones. Because the buffer advantages of wetlands derive from landscape setting rather than hydrologic properties or functions, nonwetland riparian zones which have the same landscape setting and, in all likelihood, more advantageous soil and hydrologic properties, should, if anything, have a higher priority. The general types of programs listed in Table 1 do not ordinarily distinguish between wetlands and nonwetlands though they typically include single numbers for buffer widths which are more appropriate to the better soils. Local and regional conservation and land use plans, however, often "redline" wetlands without considering vegetated nonwetland riparian areas.

Second, in many situations wetlands should not be relied upon as buffers. Rather, wetlands themselves should be buffered. Few wetlands are "sponges" for runoff — they are more likely to be runoff-generation sources and are typically ineffective compared to vegetated uplands in detaining runoff or effluent, though they may be quite efficient buffers for ground water flow, especially where anaerobic conditions are desired. Given these properties and the high biological values of many wetlands, it does not make sense to rely on riparian wetlands to filter surface water pollution from adjacent land. Where the adjacent land use is intensive (urban/residential/industrial, chemical-intensive agriculture, high-density grazing, waste applications, etc.) the riparian wetland should itself be buffered from runoff by a vegetated filter strip. This issue is addressed by shoreline or riparian management programs which use wetland/upland contacts rather than water lines as a baseline. It is also addressed by multizoned riparian corridor schemes such as those developed by the U.S. Forest Service (1992) and the State of Kansas (Dodd *et al.*, 1993). In these schemes the immediate streamside (wetland?) zone is left undisturbed and is buffered by a forested strip where occasional harvesting to remove sequestered nutrients and other contaminants may occur. This in turn is buffered by a vegetated (but not necessarily forested) low-intensity-use strip where flow is supposed to be deconcentrated.

Finally, the need to preserve and manage naturally vegetated uplands should be further investigated. Upland ecosystems are not systematically protected under any regulatory programs and thus lack a political constituency. The hydrologic values of such areas, in terms of water quality and reduction of runoff peaks, are likely to be higher, on average, than those of wetlands. Expanding regulatory protection to uplands is unlikely at best. But there are bound to be other means for encouraging the protection of such areas to provide water quality benefits.

Another likely reason for natural resource managers' attention to wetlands and relative inattention to uplands is that in many landscapes relatively pristine examples of wetland ecosystems are apparent and, by contrast, readily draw attention to wetland destruction and modification. There are, however, few remaining pristine upland systems. For example, in three coastal counties of eastern North Carolina (Craven, Carteret, and Pamlico), Moorhead and Cook (1992) estimate that 44% of the wetlands which existed at the time of European contact have been lost. While this is an appalling statistic, 56% of the original wetlands remain and some are in relatively pristine condition.

Analagous methods were used to estimate the fate of mesic upland hardwood forests in the same three counties (Phillips, 1994). More than 99% of this ecosystem has been lost, with only two isolated tracts of less than 1 ha each remaining in pristine condition. Even if all mesic uplands now succeeding from pine to oaks were undisturbed indefinitely, only about 26% of the pre-European hardwoods would have been restored a century from now. The situation is probably at least broadly similar elsewhere in North America. Brinson (1993) has already pointed out that there is not necessarily a systematic relationship between wetland values and hydroperiod. This discussion takes that argument a step farther, into the uplands.

CONCLUSIONS

While the hydrologic buffer values of wetlands have been widely touted, a realistic appraisal must recognize two facts. First, other things being equal, nonwetland environments are typically superior filters for runoff, particularly stormwater, than wetlands. Second, wetlands derive their buffer values mainly from their landscape settings, not from their hydrologic roles or properties. Thus, vegetated riparian buffer zones should be managed without regard to their wetland status. Further, many wetlands should themselves be buffered from intensive land uses and their associated runoff or effluent. Finally, the conservation of vegetated upland areas deserves further consideration.

Filtering runoff is only one of the hydrologic buffer roles of wetlands, and hydrologic buffering is only one of many wetland functions and values. The limitations of wetlands as buffers and the need to buffer riparian wetlands themselves should be considered in assessing wetlands. The superior hydrologic filter properties of many nonwetlands should also be considered in managing vegetated riparian zones and corridors.

REFERENCES

Brinson, M. M., 1993. Changes in the functioning of wetlands along environmental gradients. *Wetlands* 13:65–74.

Beven, K. J., 1986. Hillslope runoff processes and flood frequency characteristics. in Abrahams A. D., ed., *Hillslope Processes* Boston: Allen and Unwin, pp. 187–202.

Beven, K. J. and Kirkby, M. J., 1979. A physically based variable contributing area model of basin hydrology. *Hydrological Sciences Bulletin* 24:43–69.

Budd, W. W., Cohen, P. L., Saunders, P. R., and Steiner, F. R., 1987. Stream corridor management in the Pacific Northwest. *Environmental Management* 11:587–605.

Copeland, B. J. and Riggs, S. R., 1984. The Ecology of the Pamlico River, North Carolina: An Estuarine Profile. Washington D.C., *U.S. Fish and Wildlife Service* FWS/OBS-82/06.

Daniels, R. B., Kleiss, H. J., Buol, S. W., Byrd, H. J., and Phillips, J. A., 1984. Soil Systems of North Carolina. Raleigh: NC *Agricultural Research Bulletin* 467.

Diamond, R. S. and Nilson, D. J., 1988. Buffer delineation method for coastal wetlands in New Jersey, in Lyke, W. and Hoban, T., eds., *Coastal Water Resources* Bethesda, MD: American Water Resources Association, pp. 771–783.

Devito, K. J. and Dillon, P. J., 1993. The influence of hydrologic conditions and peat oxia on the phosphorus and nitrogen dynamics of a conifer swamp. *Water Resources Research* 29:2675–2685.

Dodd, R. C., McCarthy, M., Cooter, W. S., Wheaton, W. H., and Stichter, S., 1993. Riparian Buffers for Water Quality Enhancement in the Albemarle Pamlico Area. Raleigh: Albemarle-Pamlico Estuarine Study Report.

Magette, W. L., Brinsfield, R. B., Palmer, R. E., and Wood, J. D., 1989. Nutrient and sediment removal by vegetated filterstrips. *Transactions of the ASAE* 32:663–667.

McCuen, R. H., 1982. A Guide to Hydrologic Analysis Using SCS Methods. Englewood Cliffs, NJ: Prentice-Hall.

Moore, I. D. and Burch, G. J., 1986. Physical basis of the length-slope factor in the Universal Soil Loss Equation. *Soil Science Society of America Journal* 50:1294–1298.

Moorhead, K. K. and Cook, A. E., 1992. A comparison of hydric soils, wetlands and land use in coastal North Carolina. *Wetlands* 12:99–105.

Nieswand, G. H., Hordon, R. M., Shelton, T. B., Chavooshian, B. B., and Blarr, S., 1990. Buffer strips to protect water-supply reservoirs: A model and recommendations. *Water Resources Bulletin* 26: 959–966.

O'Loughlin, E. M., 1986. Partition of surface saturation zones in natural catchments by topographic analysis. *Water Resources Research* 22:794–804.

Phillips, J. D., 1989a. An evaluation of the factors determining the effectiveness of water quality buffer zones. *Journal of Hydrology* 107:133–145.

Phillips, J. D., 1989b. Nonpoint source pollution control effectiveness of riparian forests along a coastal plain river. *Journal of Hydrology* 110:221–237.

Phillips, J. D., 1989c. Evaluation of North Carolina's estuarine shoreline area of environmental concern from a water quality perspective. *Coastal Management* 17:103–117.

Phillips, J. D., 1989d. Effect of buffer zones on estuarine and riparian land use in eastern North Carolina. *Southeastern Geographer* 29:136–149.

Phillips, J. D., 1989e. Evaluating estuarine shoreline buffer zones for nonpoint source pollution control, in Magoon, O. T., ed., *Coastal Zone '89* New York: American Society of Civil Engineers, pp. 399–411.

Phillips, J. D., 1990. A saturation-based model for wetland identification. *Water Resources Bulletin* 26:333–342.

Phillips, J. D., 1994. The forgotten hardwoods of the coastal plain. *Geographical Review* 84:162–171.

Phillips, L. R. and Phillips, J. D., 1988. Land use planning techniques for estuarine shoreline buffer zone establishment, in Lyke, W. and Hoban, T., eds., *Coastal Water Resources* Bethesda, MD: American Water Resources Association, pp. 635–640.

PTRF (Pamlico-Tar River Foundation), 1991. A River of Opportunity: A Pollution Abatement and Natural Resource Management Plan for the Tar-Pamlico Basin. Washington, NC: PTRF.

Richards, K., 1982. *Rivers: Form and Process in Alluvial Channels* London: Methuen.

SCS (Soil Conservation Service), 1985. *National Soils Handbook* Washington, D.C., U.S. Department of Agriculture.

SCS, 1987. *Hydric Soils of the United States*. Washington, D.C., U.S. Department of Agriculture.

Shaw, E., 1988. *Hydrology in Practice* New York: Van Nostrand Reinhold.

Troendle, C. A., 1985. Variable source area models, in Anderson, M. G. and Burt, T. P., eds., *Hydrological Forecasting* New York: John Wiley, pp. 347–403.

USFS (U.S. Forest Service), 1992. Riparian Forest Buffers: Function and Design for Enhancement of Water Resources, Radnor, PA: USFS.

15

EFFECT OF BUFFER STRIPS ON CONTROLLING SOIL EROSION AND NUTRIENT LOSSES IN SOUTHERN FINLAND

J. Uusi-Kämppä and T. Yläranta

ABSTRACT

Agriculture is the main activity that is responsible for contributing large quantities of phosphorus and nitrogen to aquatic ecosystems in Finland. Therefore, it is becoming more important to develop ways of decreasing nutrient losses from agricultural areas by modifying cultivation practices and establishing vegetated buffer strips between agricultural fields and watercourses. Experimental plots were set up at Jokioinen in southern Finland to study the water quality improvement function of grass buffer strips (GBS) and vegetated buffer strips (VBS) in reducing soil and nutrient losses. The effectiveness of the buffer strips was studied by comparing loss of nutrients and total solids from planted areas having no buffer strips (NBS). The total solids lost from the GBS (1100 kg/ha) and VBS (900 kg/ha) were less than the total solids lost from the areas with NBS (1500 kg/ha). The sediment-adsorbed phosphorus loadings were 20 and 36% lower where the GBS and VBS, respectively, were used compared to the loading in the NBS plots. However, the loss of orthophosphates in the spring snowmelt was very high from the plots with VBS as compared to the plots with GBS and NBS. The nitrogen loss from the GBS (6 kg N/ha) was nearly half that with NBS. Buffer strips with dense vegetation can be effective in minimizing pollution of watercourses in agricultural areas.

INTRODUCTION

Finland lies between 60° and 70° north latitude and is located at the same latitude as Alaska in North America and Siberia in Asia. The surface area of

1-56670-147-3/96/$0.00+$.50
© 1996 by CRC Press, Inc.

Finland is 338,000 km^2 of which about 7.2% (24,410 km^2) is used for agriculture. One third of all agricultural land is used for the production of hay (*Phleum pratense* L., *Festuca pratensis* Huds., *Trifolium pratense* L.) to support livestock (Agricultural Information Centre, 1991).

In the southwest part of the country, where the soils are predominantly clay, more than 30% of the land is used for agriculture. In central Finland the soils are mostly silt and fine sand, and in the north, peat and organic rich mineral soils are usual. The main crops cultivated in the south and southwest parts of Finland are cereal grains. The use of the land for hay production has decreased significantly in the last few decades and more cereal crops are now cultivated. The standard technique in Finland for preparing land for planting cereal is moldboard plowing in the autumn and harrowing in early spring before sowing. Annual phosphorus and nitrogen fertilization for cereal grains in Finland is approximately 20 and 100 kg/ha, respectively (Agricultural Information Center, 1993).

The movement of nitrogen and phosphorus has been measured in selected drainage basins with different soil types and land uses in Finland (e.g., Särkkä, 1972; Kauppi, 1978, 1979a,b; Rekolainen, 1989, 1993). Rekolainen (1989) found the annual loss of nutrients to be 0.9–1.8 kg P/ha and 8–20 kg N/ha between 1981 and 1985. It is estimated that the total annual phosphorus loss from cultivated fields for all of Finland is between 2000–4000 tons, and between 20,000–40,000 tons of nitrogen (Rekolainen, 1989). Jaakkola (1984) and Turtola (1993) have been studying nutrient loss from different cropping systems subject to different fertilization and cultivation techniques in a field test site at the Agricultural Research Centre at Jokioinen since 1975. The annual loss was 12 kg N/ha and 1.1 kg P/ha from fields of spring barley grown on clay soil (Turtola and Jaakkola, 1985). Mansikkaniemi (1982) studied fluvial systems in southwestern Finland and found that 500 kg/ha of total solids were lost per year from certain catchment areas. During years of heavy rainfall, as much as 5000–6000 kg/ha total solids were lost from agricultural fields into the watercourses. For a better understanding of the process, comparable field experiments have been established elsewhere in Finland on different soils such as sand, silt, and peat. At Jokioinen, lysimeters have been in place since 1981 to examine nutrient loss from different soils (Jaakkola, 1984; Yläranta et al., 1993; Turtola, 1993), and data collection and analyses are ongoing.

Eutrophication of lakes, rivers, and Baltic Sea coastal waters by high phosphorus and nitrogen inputs is one of the most serious water pollution problems facing Finland (Kauppi, 1993). Experimental results have shown that 75% of the phosphorus transported to watercourses is bound to sediment and about 5% of the sediment-bound phosphorus is bioavailable for algal growth (Ekholm et al., 1991). A total of 29% of the phosphorus (soluble and adsorbed) lost is biologically available for aquatic plants (Ekholm, 1992). It has also been pointed out that phosphorus and nitrogen losses from agricultural land are higher compared to those generated from industrial and municipal loads (Rekolainen et al., 1992). Due to these considerations, the environmental policy in Finland has been targeted to reduce inputs of pollutants, especially phosphorus and nitrogen, released to surface waters from all sources by 1995 (Ministry of Environment, 1988). The strategy calls for reduction of phosphorus losses by 30% and a significant reduction in nitrogen loading from agricultural fields by 1995. The Finnish agencies are embarking on several research programs to develop better management options for agriculture.

It is now known that nitrogen losses from the land can be reduced by the use of crops such as hay (*Phleum pratense* L. and *Festuca pratensis* Huds.) or catch crops with long growing periods following harvesting of the main crops. Changes in tillage techniques or postponement of plowing to the late autumn or early spring are also possible ways of reducing nitrogen losses (Rekolainen *et al.*, 1992). The loss of total phosphorus can be effectively reduced by decreasing the use of phosphorus fertilizers and by controlling soil erosion and subsequent transport of soil to watercourses.

The Water Protection Association of the Vantaa River and Helsinki District developed a project in the Vantaa River basin in southern Finland to study the effects of vegetative buffer strips to reduce nutrient losses from agricultural fields. According to Ahola (1989, 1990) the width of the vegetated buffer zone is determined by examining factors such as the size of the watercourse, topography of the watershed, soils, and vegetation cover.

This chapter describes the findings of an experimental field study at the Agricultural Research Centre of Finland aimed to assess the effectiveness of buffer strips to control the transport of sediment and nutrients such as phosphorus and nitrogen.

MATERIALS AND METHODS

The study site is a 6-plot experimental field (3 treatments × 2 replicates; Uusi-Kämppä and Yläranta, 1992) established in 1989 at the Agricultural Research Centre at Jokioinen in southern Finland (60°48' N and 23°28' E). The top soil has about 60% clay content (Table 1).

In 1990, the year preceding the buffer strip experiments, spring barley (*Hordeum vulgare* L.) was sown in the field and runoff was measured and water and sediment samples were taken for chemical analyses. Runoff characteristics and loss of total solids and nutrients were measured. From the measurements it was found that the values varied from plot to plot across the experimental field during the course of the year. This information was used to determine the layout of the buffer strips. Plots 2 and 5, with the lowest mean runoff and sediment yield, were selected for the control, i.e., no buffer strips. Plots 1 and 4 and plots 3 and 6, with equal mean runoff and sediment yield, were selected for the grass buffer strips and vegetated buffer strips, respectively. The measurements from each pair of plots were averaged to indicate the runoff in millimeters and nutrient and sediment losses in kilograms per hectare (Table 2).

The buffer strip experiments (Table 3) began in the autumn of 1991 to measure the effects of grass buffer strips (GBS) and vegetated buffer strips (VBS)

Table 1   Characteristics of Soils in Field and Buffer Areas of the Study Site

Characteristic	Depth (cm)	Field	Buffer area
Org. C (%)	0–20	3.0	2.3
	20–50		0.9
pH (H$_2$O)	0–20	5.9	6.2
	20–50		6.4
P (mg/l)	0–20	7.7	5.4
	20–50		1.7
Particle size distribution (%)			
<0.002 mm	0–20	56	60
0.002–0.02 mm	0–20	26	27
0.02–0.2 mm	0–20	12	11
0.2–2.0 mm	0–20	6	2

Table 2   Runoff (Surface + Subsurface), Total Solids (TS), and Plant Nutrient Losses From NBS, GBS, and VBS Plots in Pre-experimental and Experimental Periods, Mean of Two Replicates

Period	Treatment	Runoff, (mm)	TS	Tot-N	NO$_3^-$-N	NH$_4^+$-N	Part.-P	PO$_4^{3-}$-P
Pre-experimental								
Autumn 1990		15	100	3.0	2.8	0.02	0.08	0.02
Spring 1991	NBS	39	170	2.9	2.5	0.07	0.08	0.04
Total		54	270	5.9	5.3	0.09	0.16	0.06
Pre-experimental								
Autumn 1990		19	150	3.5	3.2	0.03	0.12	0.02
Spring 1991	GBS	44	220	2.8	2.4	0.09	0.11	0.04
Total		63	370	6.3	5.6	0.12	0.23	0.06
Pre-experimental								
Autumn 1990		19	140	3.8	3.5	0.03	0.12	0.03
Spring 1991	VBS	41	220	3.1	2.7	0.07	0.11	0.06
Total		60	360	6.9	6.2	0.10	0.23	0.09
Experimental								
Autumn 1991		16	330	2.0	1.4	0.02	0.27	0.02
Spring 1992		42	190	2.5	2.0	0.06	0.14	0.06
Autumn 1992	NBS	24	210	2.9	2.4	0.04	0.20	0.04
Spring 1993		64	720	5.6	4.3	0.10	0.50	0.10
Autumn 1993		10	90	0.5	0.3	0.01	0.09	0.02
Total		156	1540	13.5	10.4	0.23	1.20	0.24
Experimental								
Autumn 1991		15	170	0.5	0.3	0.01	0.18	0.02
Spring 1992		61	240	1.8	1.4	0.10	0.18	0.09
Autumn 1992	GBS	33	190	1.0	0.6	0.04	0.19	0.04
Spring 1993		56	470	2.5	1.6	0.10	0.33	0.08
Autumn 1993		18	70	0.4	0.2	0.02	0.08	0.03
Total		183	1140	6.2	4.1	0.27	0.96	0.26
Experimental								
Autumn 1991		18	200	1.2	0.7	0.02	0.18	0.03
Spring 1992		67	240	3.2	2.5	0.19	0.17	0.12
Autumn 1992	VBS	25	100	1.0	0.7	0.03	0.13	0.04
Spring 1993		62	300	3.8	2.7	0.18	0.24	0.13
Autumn 1993		11	40	0.3	0.1	0.03	0.05	0.07
Total		183	880	9.5	6.7	0.45	0.77	0.39

Table 3 Cover Types in Buffer Strip and Plot Treatments During the 1990–93 Study Period

	1990	1991	1992	1993
	Plot area			
	Spring Barley	Spring Barley	Spring Barley	Spring Oats
	Plant: 4.5	Plant: 10.5	Plant: 15.5	Plant: 12.5
	Harvest: 21.8	Harvest: 18.8	Harvest: 3.9	Harvest: 16.9
	Plow: 17.10	Plow: 17.9	Plow: 6.10	Plow: 12.10
Buffer treatment	Buffer area			
NBS	Bare	Barley	Barley	Oats
(No Buffer Strip)				
GBS	Bare	Perennial grass and barley without fertilization Planted 27.5	Mowing and removing grass cover 16.9	Mowing and removing grass cover 9.8
(Grass Buffer Strip)				
VBS	Bare	Perennial grass Planted 25.6 Trees and shrubs Planted 3.10	Grass, trees and shrubs	Grass, trees and, shrubs
(Vegetated Buffer Strip)				

on sediment and nutrient losses. Spring barley or spring oats (*Avena sativa* L.) were grown on each plot in any one year. At the lower end of each plot a 10-m-wide strip with a slope of over 10% was established. However, in the controls (NBS) barley or oats were grown on the buffer strips. On the GBS treatment plots, a mixture of timothy grass (*Phleum pratense* L.) and meadow fescue (*Festuca pratensis* Huds.) was used and the grass was mowed and removed once each growing season. In the VBS treatment plots a mixture of common bent (*Agrostis tenuis* Sibth.) and yarrow (*Achillea millefolium* L.) was grown. Birch (*Betula* L.), goat willow (*Salix caprea* L.), mountain currant (*Ribes alpinum* L.), viburnum (*Viburnum opulus* L.), alder (*Alnus* B. Ehrh.), maple (*Acer* L.), and rowan (*Sorbus aucuparia* L.) were planted in October 1991. Trees and shrubs were small and they did not have any significant effect on runoff characteristics nor on nutrient-removing capability.

Surface and subsurface water was directed into a collector ditch to a depth of 30 cm on each plot. The total volume of runoff water was measured volumetrically with a tipping bucket gauge and the number of buckets removed was continuously recorded on a clock-driven chart. A representative flow-weighted subsample was taken for analysis in the laboratory. Although the drain tubes were installed approximately 1 m below the soil surface, drainage was not measured.

Water samples (500 ml) were collected in LDPE bottles once a day during the peak runoff period or every second week during the non-peak runoff period. Total solids and concentrations of total nitrogen and total phosphorus were determined on unfiltered subsamples. For determination of orthophosphate, ammonium nitrogen, and nitrate nitrogen, samples were filtered through a membrane filter (Sartorius 11306-50PFN, pore size 0.45 µm). Total solids were determined as evaporated residue after drying at 105°C and this measurement as dry weight was taken to represent the amount of soil erosion. Total phosphorus, soluble orthophosphate

Table 4   Dry Matter Yield of Spring Barley (Grain and Straw) and Grass (kg/ha) and Plant Nutrient Uptake (kg/ha) on Field and Buffer Area in 1992

Plot area		Buffer area					
		NBS		GBS		VBS	
Plot no.	1–6	2	5	1	4	3	6
Yield							
Barley	7900	4600	3800				
Grass				2500	2500	2500	2000
Uptake							
Nitrogen	134	90	80	38	34	60	45
Phosphorus	16	11	7	5	5	8	6

phosphorus, total nitrogen, nitrate nitrogen, and ammonium nitrogen were analyzed according to the Finnish standard methods (Suomen standardisoimisliitto, 1976, 1986a,b, 1990a,b) and used Flow Injection Analysis on a LACHAT QuikChem analyzer (Lachat Instruments, U.S.). Concentration of sediment-adsorbed phosphorus was computed as the difference between total phosphorus and orthophosphate phosphorus concentrations.

The barley yield of cereal grains and straw in the plots and grass yields in the buffer strips were also measured in 1992. Nitrogen and phosphorus uptake on the plots and the buffer strips were estimated from the analyses of total nitrogen and phosphorus in the grain, straw, and grass.

RESULTS

Crop Yields

The yields of barley and the uptake of nitrogen and phosphorus were higher in the field plots compared to the barley grown on the lower end of the slope of the NBS plots (Table 4). The lower yield on the slope was obviously due to drier growing conditions. The grass production in the GBS and VBS buffer strips was comparable, but the total uptake of nitrogen and phosphorus was greater in VBS than in GBS as measured in September 1992.

Runoff and Total Solids

The amount of precipitation and total runoff from NBS during peak periods in the autumn and spring are given in Table 5. The runoff from the NBS plots was almost the same for both the pre-experimental year (1990–91) and the first year of the experiment (1991–92). In year two of the experiment (1992–93), the runoff from the NBS plots was almost 60% greater than in the previous years (Figure 1). The spring runoff in 1992 and 1993 from the GBS and VBS plots was more than 30 and 50%, respectively, than in the spring of 1991. In August 1993, the wettest month oats were grown in the field, the runoff was low during the autumn because of unusually low rainfall. During the pre-experimental period, runoff from the VBS and GBS plots was 11–16% greater than from the NBS plots and the relative

Table 5 Precipitation (mm) and Runoff (mm) From NBS in Pre-experimental and Experimental Periods

		Runoff	
Period	Precipitation (mm)	Mean (mm)	Range (mm)
Pre-experimental			
Summer 1990 (4.5.-2.8.1990)	127		
Autumn 1990 (2.8.-31.12.1990)	315	15	13–17
Spring 1991 (1.1.-16.5.1991)	145	39	33–45
Total	587	54	46–62
Experimental			
Summer 1991 (17.5.-24.9.1991)	280		
Autumn 1991 (25.9.-31.12.1991)	188	16	14–18
Spring 1992 (1.1.-21.4.1992)	167	42	40–44
Total	635	58	54–62
Summer 1992 (22.4.-5.10.1992)	251		
Autumn 1992 (6.10.-31.12.1992)	160	24	18–30
Spring 1993 (1.1.-26.4.1993)	130	64	47–81
Total	541	88	65–111
Summer 1993 (27.4.-10.8.1993)	207		
Autumn 1993 (11.8.-31.12.1993)	220	10	4–16

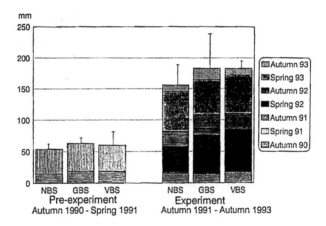

Figure 1   Runoff (surface and subsurface) from the NBS, GBS, and VBS plots in pre-experimental and experimental periods. Mean of two replicates with error bar showing maximum value.

difference remained approximately 17% greater than the NBS plots during the same experimental period. The runoff measured during this study, in 1990–91 and 1991–92, was three to four times less than values reported by Puustinen (1993) for fields planted in winter wheat on heavy clay soil in southwestern Finland.

The loss of total solids expressed as the sum of five runoff periods from the plots with GBS and VBS were 74 and 57%, respectively, of the NBS plots (Figure 2). In the autumn of year one, short grass growth and straw of barley on the GBS plots reduced total solids in the runoff water by an average of 160 kg/ha (48%). However, in the spring of 1992, runoff from the GBS and VBS plots averaged

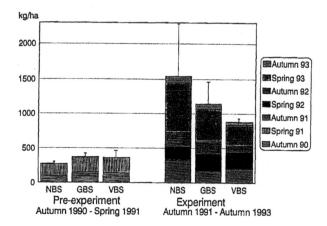

Figure 2   Erosion from the NBS, GBS, and VBS plots in pre-experimental and experimental periods. Mean of two replicates with error bar showing maximum value.

about 50% more than that from the NBS plots. Therefore, the losses of suspended solids from the GBS and VBS plots were 240 kg/ha, 26% greater than from the NBS plots. This increased sediment yield from the GBS and VBS in spring 1992 resulted from grass removal on the GBS plots and because most of the cover on the VBS plots was from the annual grasses of autumn 1991.

In the autumn of year two, the harvest of the grass in late September on the GBS plots only decreased the loss of total solids by 9%. On the VBS plots, dense vegetation of mainly common bent and yarrow grasses decreased the loss of total solids by 57% (530 kg/ha) compared to the loss from the NBS in year two.

Phosphorus Losses

In the pre-experimental year, the loss of sediment-adsorbed phosphorus from the buffer strip plots was over 40% greater than from the NBS plots. The GBS and VBS reduced the loss of sediment-adsorbed phosphorus during the experimental years by 20 and 36%, respectively, compared to the NBS plots (Figure 3).

In the autumn of year one, the buffer strips decreased the loss of sediment-adsorbed phosphorus by 0.09 kg/ha (33%) compared to the NBS plots. In the spring of year one, the buffer strips were poorly covered and therefore had about 20% higher phosphorus loss than the NBS plots. In the autumn of year two, the grass was harvested and removed late in September and the relatively little remaining grass did not act as a good buffer for phosphorus reduction on the GBS. However, in the spring of year two, GBS reduced sediment-adsorbed phosphorus loss by 34% (0.17 kg/ha). The VBS had dense vegetation which reduced loss of sediment-adsorbed phosphorus by 0.33 kg/ha (47%) in year two. By the autumn of year three, the loss of sediment-adsorbed phosphorus was small (0.05–0.09 kg/ha).

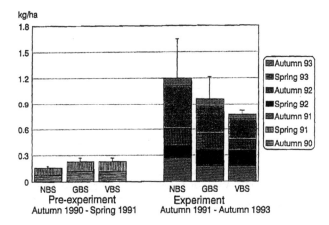

Figure 3   The loss of sediment-adsorbed phosphorus from the NBS, GBS, and VBS plots in pre-experimental and experimental periods. Mean of two replicates with error bar showing maximum value.

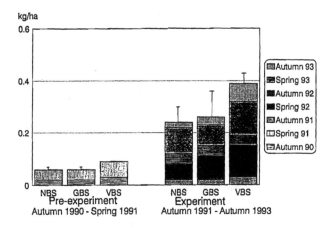

Figure 4   The loss of orthophosphate phosphorus from the NBS, GBS, and VBS plots in pre-experimental and experimental periods. Mean of two replicates with error bar showing maximum value.

The loss of soluble orthophosphates from the VBS in the pre-experimental year was 50% (0.03 kg/ha) greater than from the NBS and GBS plots (Figure 4). The loss of orthophosphates from VBS continued to be greater than from the other plots in the experimental years. This means that the VBS did not reduce soluble phosphorus.

In the experiment, the concentration of orthophosphates did not vary much between the NBS and GBS plots (Figure 5). On the VBS plots the concentration of soluble phosphorus was slightly greater than in the other plots. However, in the autumn of 1993 the concentration was high in the VBS.

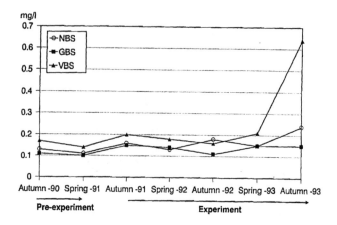

Figure 5   Concentration of orthophosphate phosphorus in runoff water from the NBS, GBS, and VBS plots in pre-experimental and experimental periods. Mean of two replicates.

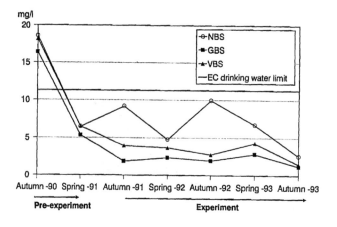

Figure 6   Concentration of nitrate nitrogen in runoff water from the NBS, GBS, and VBS plots in pre-experimental and experimental periods. Mean of two replicates.

Nitrogen Losses

About 90% of the total nitrogen was in the form of nitrate in the pre-experimental year. However, during years one and two of the experiment only about 60–80% of the nitrogen was available in the form of nitrate nitrogen. Due to dry conditions in the autumn of year three, only about 30–60% existed as nitrate nitrogen. In the autumn of the pre-experimental year, the concentration of nitrate nitrogen in runoff water exceeded the EEC drinking water limit (EEC, 1980). The concentrations were less (Figure 6) during the experimental years, but in plot 5 (NBS) the concentration did exceed the limit in the autumn of years one and two.

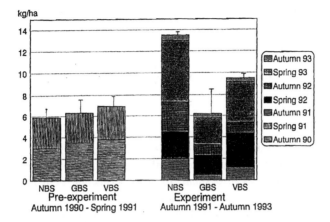

Figure 7 Total nitrogen from the NBS, GBS, and VBS plots in pre-experimental and experimental periods. Mean of two replicates with error bar showing maximum value.

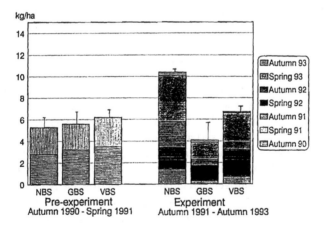

Figure 8 Nitrate nitrogen in runoff from the NBS, GBS, and VBS plots in pre-experimental and experimental periods. Mean of two replicates with error bar showing maximum value.

   The losses in runoff of total nitrogen and nitrate nitrogen were greatest in the NBS plots (Figures 7 and 8). In years one and two of the experiment, the GBS reduced the losses of total nitrogen by 49 and 59%, respectively. In the spring of year one, the vegetation growth was poor in the VBS and so nitrogen losses from the VBS were even greater than from the other plots. Nitrogen loss had also decreased in the VBS in year two.

   Nitrate nitrogen losses were also lower in the GBS and VBS plots than in the NBS plots. The losses of nitrate nitrogen from the GBS and VBS were 30 and 50%, respectively, of that from the NBS during the autumn of the first two experimental years. During the study, nitrate nitrogen losses were higher in the spring periods than in the autumn periods. Losses were also greater from the VBS than from the GBS in the spring.

The loss of ammonium nitrogen was small on the NBS, GBS, and VBS during the first two years of the experiment. The spring loss of ammonium nitrogen was greater than the autumn loss in all treatments (Table 2).

DISCUSSION

Runoff was generally low in the experimental plots, even including values for shallow subsurface runoff. Consequently the loss of total solids and sediment-adsorbed phosphorus was also small. In an experimental field near Turku in southwestern Finland, Puustinen (1993) found runoff rates three to four times and loss of total solids three times greater than those observed in this study. Puustinen (1993) used winter wheat (*Triticum aestivum* L.). Mansikkaniemi (1982) studied the quality of water in several catchment streams in southwestern Finland and found ten times more sediment (5000–6000 kg/ha) removed from the sloping fields compared to what is reported in this study in the wettest years. The small loss of total solids from field sites in this study is attributed to a much gentler slope and much less runoff.

The losses of sediment-adsorbed phosphorus from the NBS were also quite small compared to other data from southwestern Finland. Seven times more sediment-adsorbed phosphorus has been found in surface and subsurface waters on clay loam soils in a field only 70 km southwest of Jokioinen (Turtola and Joakkola, 1985; Puustinen, 1993). Turtola and Jaakkola (1985) recorded slightly higher losses of phosphorus in surface waters from an experimental field of barley also located at Jokioinen. The loss of soluble orthophosphates in this study was about 25% of that from plots sown in winter wheat (Puustinen, 1993).

The losses of total nitrogen and nitrate nitrogen from the NBS were about the same as those recorded from fields of barley or winter wheat (Turtola and Joakkola, 1985; Puustinen, 1993), and the values for loss of nitrate nitrogen from buffer strip plots compared well with those found from grassed areas elsewhere in the region (Turtola and Jaakkola, 1985).

In comparing data from the present study to research elsewhere, Dillaha *et al.* (1989) in the U.S. found that 9.1-m-wide VBS on 11–16% slopes decreased losses of sediment and total phosphorus from bare cropland by an average of 84 and 79%, respectively. Similarly, Magette *et al.* (1987) reported 9.2-m-wide VBS cut losses of sediment and total phosphorus in fallow fields in the U.S. by 86 and 53%, respectively. The data for year two presented in this study on the effectiveness of the VBS in reducing losses of sediment and sediment-adsorbed phosphorus compare well with those of Magette *et al.* (1987). Unfertilized GBS in our study were not nearly as effective in minimizing sediment transport and the loss of sediment-bound phosphorus as the VBS. It was not until the second year of the experiment that vegetation in both the GBS (59%) and VBS (43%)

was established well enough to reduce losses of total nitrogen closer to values observed in the U.S.

Decreases in the losses of total solids and sediment-adsorbed phosphorus were slightly less in our study compared to those in the U.S. (Dillaha *et al.*, 1989; Magette *et al.*, 1987). There are four possible reasons for this. First, this field study was carried out without irrigation, whereas a rainfall simulator was used to apply 100 mm of water to each plot over a two-day period at the study sites in the U.S. Second, the climate of southwestern Finland is quite different; snow covers the fields in winter and there is no growth of vegetation during the main runoff period of early spring. The winters, however, are mild and vegetation grows year round at the study sites in the U.S. Third, the sites in the U.S. were fertilized and were left bare before rainfall simulations, whereas in our experiment the fields were sown with barley which took nutrients out of the soil during the summer. Finally, the soils at Jokioinen are clay rich and much finer in texture compared to the sites in the U.S.

The leaching of soluble phosphates showed a slight increase from the VBS in spring. Dillaha *et al.* (1989) found that phosphorus yields from the buffer strips were larger than the inputs to the buffers. They indicated that there was a tendency for the phosphorus trapped earlier to be released as soluble phosphorus from the buffer strip vegetation and soil.

CONCLUSIONS

The following conclusions and observations can be made from this study:

- In the VBS, dense vegetation of grasses decreased the losses of total solids and sediment-adsorbed phosphorus significantly.
- In the GBS, the mowing and removal of grass late in the growing period significantly decreased losses of total solids, especially sediment-adsorbed phosphorus, nitrate, and total nitrogen.
- The loss of soluble phosphorus was over 50% greater from the plots with VBS as compared to the plots with NBS and GBS.
- Higher losses of orthophosphates from the VBS plots compared to the other plots, and higher losses of total and nitrate nitrogen from the VBS plots compared to the GBS plots, were likely because nutrients were leached from the decaying grass residue on the surface of the buffer strip in the spring.
- It is concluded that the grass buffer strips (GBS) give reasonable reductions of sediment yield and nutrient losses from crops on clay soils in southwestern Finland. Only sediment yield-associated adsorbed phosphorus can be further reduced with the vegetated buffer strips (VBS). However, to maintain reductions in orthophosphates losses, the vegetation on buffer strips or buffer zones should be mowed and the mowings removed.
- The results presented in this study are preliminary and need further confirmation. The study has been conducted on experimental plots and should be extended to field-size areas. Further, the empirical plot and field data should be validated with mathematical models.

REFERENCES

Agricultural Information Center, 1991. *Northern Vitality Agriculture in Finland*, 21.

Agricultural Information Center, 1993. *Agrifacts '93 about Finland*, 25.

Ahola, H., 1989. Vegetated buffer zone project of the Vantaa River river basin, *Geografisk Tidsskrift*, 89:22–25.

Ahola, H., 1990. Vegetated buffer zone examinations on the Vantaa River basin. *Aqua Fennica*, 20:65–69.

Dillaha, T. A., Reneau, R. B., Mostaghimi, S., and Lee, D., 1989. Vegetative filter strips for agricultural nonpoint source pollution control, *Transactions of the ASAE*, 32:513–519.

Ekholm, P., 1992. Maataloudesta peräisin oleva fosfori vesien rehevöittäjänä, in Rekolainen, S. and Kauppi, L., eds., *Maatalous ja vesien kuormitus*, Yhteistutkimusprojektin tutkimusraportit, Vesi- ja ympäristöhallituksen monistesarja No. 359, Helsinki, 39–46 (In Finnish).

Ekholm, P., Yli-Halla, M., and Kylmälä, P., 1991. Availability of phosphorus in suspended sediments estimated by chemical extraction and bioassay. *Verhandlungen der Internationale Vereinigung für theoretische und angewandte Limnologie*, 24:2994–2998.

European Economic Community (EEC) L229, 1980. Council directive on the quality of water for human consumption, *Official Journal*, No. 80/778, 11.

Jaakkola, A., 1984. Leaching losses of nitrogen from a clay soil under grass and cereal crops in Finland, *Plant and Soil*, 76:59–66.

Kauppi, L., 1978. Effects of drainage basin characteristics on the diffuse load of phosphorus and nitrogen, *Publications of the Water Research Institute*, National Board of Waters, Finland, 30, 21–41.

Kauppi, L., 1979a. Effects of land use on the diffuse load of phosphorus and nitrogen. *Nordic Hydrology*, 10:79–88.

Kauppi, L., 1979b. Phosphorus and nitrogen input from rural population, agriculture and forest fertilization to watercourses, *Publications of the Water Research Institute*, National Board of Waters, Finland, 34, 35–46.

Kauppi, L., 1993. Contribution of agriculture to nutrient loading of surface waters in Nordic countries, in Elonen, P. and Pitkänen, J., eds., *Soil Tillage and Environment*, NJF-Utredning/Rapport NR. 88, Proceedings of NJF-seminar no. 228, Scandinavian Association of Agricultural Scientists (NJF), Jokioinen, Finland, 1–5.

Magette, W. L., Brinsfield, R. B., Palmer, R. E., Wood, J. D., Dillaha, T. A., and Reneau, R. B., 1987. *Vegetated Filter Strips for Agricultural Runoff Treatment*, CBP/TRS 2/87, U.S. Environmental Protection Agency, Philadelphia, PA 19107, 125.

Mansikkaniemi, H., 1982. Soil erosion in areas of intensive cultivation in southwestern Finland. *Fennia*, 160:225–276.

Ministry of Environment, 1988. *Water Protection Program until 1995 in Finland*, Series B, 12/1988, Helsinki, 40.

Puustinen, M., 1993. Effect of soil tillage on erosion and nutrient transport in surface runoff, in Elonen, P. and Pitkänen, J., eds., *Soil Tillage and Environment*, NJF-Utredning/Rapport NR. 88, Proceedings of NJF-seminar no. 228, Scandinavian Association of Agricultural Scientists (NJF), Jokioinen, Finland, 218–224.

Rekolainen, S., 1989. Phosphorus and nitrogen load from forest and agricultural areas in Finland. *Aqua Fennica*, 19:95–107.

Rekolainen, S., 1993. Assessment and mitigation of agricultural water pollution. *Publications of the Water and Environment Research Institute*, National Board of Waters and Environment, Finland, No. 12, 34 + 8 papers.

Rekolainen, S., Kauppi, L., and Turtola, E., 1992. *Maatalous ja vesien tila, Agriculture and State of Waters*, Final report of the MAVERO-project, Ministry of Agriculture and Forestry, Luonnonvarainjulkaisuja 15, Helsinki, 61, (In Finnish with English and Swedish summary).

Suomen standardisoimisliitto, 1976. *Veden ammoniumtypen määritys.* (Finnish standard concerning determination of ammonia-nitrogen of water), SFS 1932, 6. (In Finnish).

Suomen standardisoimisliitto, 1986a. *Veden fosfaatin määritys.* (Finnish standard concerning determination of phosphate in water), SFS 3025, 10. (In Finnish).

Suomen standardisoimisliitto, 1986b. *Veden kokonaisfosforin määritys. Hajotus peroksodisulfaatilla.* (Finnish standard concerning determination of total phosphorus in water. Digestion with peroxodisulphate), SFS 3026, 11. (In Finnish).

Suomen standardisoimisliitto, 1990a. *Veden nitriitti- ja nitraattitypen summan määritys.* (Finnish standard concerning determination of the sum of nitrite and nitrate nitrogen in water), SFS 1930, 5. (In Finnish).

Suomen standardisoimisliitto, 1990b. *Veden typen määritys. Peroksodisulfaattihapetus.* (Finnish standard concerning determination of nitrogen in water. Oxidation with peroxodisulfate), SFS 1931, 6. (In Finnish).

Särkkä, M., 1972. The washing out of nutrients in the watersheds. *Aqua Fennica*, 88–103.

Turtola, E., 1993. Phosphorus and nitrogen leaching during set-aside, in Elonen, P. and Pitkänen, J., Eds., *Soil Tillage and Environment*, NJF-Utredning/Rapport NR. 88, Proceedings of NJF-seminar no. 228, Scandinavian Association of Agricultural Scientists (NJF), Jokioinen, Finland, 207–217.

Turtola, E. and Jaakkola, A., 1985. *Viljelykasvin ja lannoituksen vaikutus typen ja fosforin huuhtoutumiseen savimaasta,* Tiedote 6/85, Maatalouden tutkimuskeskus, 43. (In Finnish).

Uusi-Kämppä, J. and Yläranta, T., 1992. Reduction of sediment, phosphorus and nitrogen transport on vegetated buffer strips, *Agricultural Science in Finland*, 1:569–575.

Yläranta, T., Uusi-Kämppä, J., and Jaakkola, A., 1993. Leaching of nitrogen in barley, grass ley and fallow lysimeters, *Agricultural Science in Finland*, 2:281–291.

16

HYDROGEOLOGICAL CRITERIA FOR BUFFER ZONES BETWEEN WETLANDS AND AGGREGATE EXTRACTION SITES

J. Z. Fraser, D. Routly, A. B. F. Hinton, and K. Richardson

ABSTRACT

Although pits and quarries have been established and operated in close proximity to wetlands throughout southern Ontario, the nature and extent of their impacts on wetland processes and functions are not well documented. A better understanding of these impacts on local hydrogeological processes and wetland functions and on the environmental gradients which define wetland boundaries is required to effectively implement Ministry of Natural Resources wetlands and aggregate resource management policies.

The Southern Region Science and Technology Transfer Unit of the Ministry of Natural Resources is currently assessing the range and scale of these impacts to: identify situations where significant impacts may occur; improve siting criteria and define buffer zones; identify operating practices which mitigate negative impacts; maintain local hydrogeological processes; and to develop rehabilitation strategies for the long-term protection and enhancement of adjacent wetlands. This effort is part of long-term monitoring and experimental management initiatives which will be aimed at improving our understanding of wetland gradients, boundaries, and buffer zones in relation to aggregate extraction.

INTRODUCTION

Understanding and minimizing the impacts of aggregate extraction on local hydrogeological process and on the environmental gradients which define adjacent wetland boundaries is a major concern for both the Ontario Ministry of Natural Resources and the general public. Effective regulation of existing operations, as well as review and approval of new sites under terms of the Aggregate

1-56670-147-3/96/$0.00+$.50

Resources Act, require the development of criteria for mitigation strategies and buffer zones which will protect wetlands from these impacts.

Presently, although a large number of extractive sites in the Southern Region of Ontario are located in proximity to Provincially Significant Wetlands, the nature and extent of their impacts on environmental gradients and wetland functions are not well documented. As a result, the Science and Technology Transfer Unit in the Southern Regional Office has recently initiated an assessment of the range and scale of these impacts to: improve siting criteria and define optimum buffer zones; identify operating practices which mitigate negative impacts and maintain local hydrogeological processes; and develop rehabilitation strategies for the long-term protection and enhancement of wetlands.

In the future, effective solutions to minimize the impacts of extraction on wetlands may focus on the implementation of strategic resource management plans at a regional scale and over the long term. The current development of an integrated planning strategy for the Oak Ridges Moraine is an example of the type of comprehensive landscape ecology approach to multi-objective resource management and protection strategies which will characterize planning in the future (Geomatics International, 1993). The use of large data sets for a wide variety of resource values in a GIS mode will permit strategic modelling and assist in providing decision-support systems which will optimize extraction of aggregate resources with the least impact on sensitive environments.

MANAGING AGGREGATES AND PROTECTING WETLANDS

The extraction of mineral aggregates which include natural sand and gravel deposits, as well as crushed stone from bedrock, is an important component of the mining sector in Ontario. In 1992, the latest year for which figures are available, more than 97 million tonnes of aggregates were extracted from nearly 2400 licensed sites in the heavily populated southern part of the province (Ministry of Natural Resources, Resource Stewardship and Development Branch, 1994).

Although aggregate resources are plentiful in Southern Ontario, their distribution is irregular. They are nonrenewable resources which can only be extracted where they are geologically favorable. Options for development of alternate resource areas are further reduced by the high bulk and low unit value of the raw material. The economic viability of a deposit is thus a function of its proximity to market areas as well as its quality and size. Recent trends in aggregate utilization indicate that significant resources will be required throughout southern Ontario and especially in the Greater Toronto area (GTA) over the next 15 years (Planning Initiatives Ltd. and Associates, 1993). Because of geological and economic siting constraints, deposits suitable for large-scale development are often in conflict with other land uses.

Mitigating the impacts of extraction on water resources is a major aspect of aggregate resource management in southern Ontario. However, it is important to make a distinction between issues directly related to the protection of wetland functions and other issues which are more related to the use of ground and surface waters for domestic consumption or for other economic activities. Although both sets of issues are often interconnected, this chapter will focus on those directly related to the protection of wetlands.

Legislative and Policy Context for Aggregate Extraction

The extraction and management of aggregate resources in Ontario are regulated through the Aggregate Resources Act (RSO, 1990) and the Mineral Aggregate Resources Planning Policy (Order in Council No. 1249-86). Although both of these regulatory instruments note the importance of reducing land use conflicts and protecting surface and ground water resources, specific buffer zones or separation distances between wetlands and aggregate extraction activities are not identified. Because of the unique set of conditions which characterize each proposal for aggregate extraction, strategies for mitigation and separation are also unique and site specific. The Aggregate Resources Act recognizes this and requires a detailed site review process. Section 8 of the Act requires that site plans accompanying applications for licenses must indicate:

> "8(1)(o) the water table and any existing surface water on and surrounding the site and proposed water diversion, storage and drainage facilities on the site and points of discharge to surface waters; 8(1)(p) subject to available information, the location of water wells on and within 300 metres of the site; 8(1)(q) the maximum depth of extraction and whether it is intended to excavate below the water table."
>
> Applications must also be accompanied by a report where necessary, "9(1)(b) describing the environment that may be expected to be affected by the pit or quarry operation and any proposed remedial measures that are considered necessary."

Section 12 of the Act states that the Minister of Natural Resources must consider the effect of the pit or quarry on the environment in deciding on whether to issue a license, and more specifically must consider any possible effects on ground or surface water resources. However, the Act does not specify separation requirements or buffer zones between extractive activities and wetlands beyond a statement in Regulation 18 of the Act which states that: "No person shall excavate setback areas within fifteen metres from any water body that is not the result of excavation below the water table."

Generally, the assessment of possible impacts of extraction on water-related resources is undertaken through detailed review of information provided by Sections 8 and 9 of the Act and evaluated on a site-specific basis as part of Section 12. This review may identify approaches to mitigation of activities or development of buffer zones to protect water resources. To date, site-specific solutions have not been evaluated as a group to develop general guidelines for extractive activities.

A preliminary survey of current licenses in southern Ontario was undertaken as part of this study. Cambridge, Maple, and Aylmer, the three districts which occupy approximately half of the region and represent the most heavily populated portions of the region, were selected for review. There are a total of 1035 licenses issued in these districts. Of these sites, 141 or 14% are located within 500 m of a Provincially Significant wetland and 323 sites or 31% are located within 500 m of a perennial watercourse. A total of 401 sites, representing nearly 40% of all licenses in the three districts have extracted, or propose to extract below the local ground water level. Despite a high level of possible interaction of extraction activity with water resources, formal license conditions relating to local hydrogeology are included in only 75 sites, representing less than 7% of all licensed properties within the three districts. These statistics reflect the need for a better understanding of the impacts of extraction on water resources.

Legislative and Policy Context for Wetland Protection

The Ministry of Natural Resources has a number of water-related resource management mandates based on the administration of provincial and federal legislation, including:

Fisheries habitat management, in cooperation with the federal government in administering the Federal Fisheries Act,

Water quantity and quality management in cooperation with the Ministry of Environment and Energy and Conservation Authorities, based on the administration of the Ontario Water Resources Act, Conservation Land Act, Lakes and Rivers Improvement Act, and other related legislation.

Wetland protection is undertaken through the combined implementation of this legislation, the stewardship efforts of a variety of nongovernment agencies and private landowners, and through the implementation of the Wetlands Policy Statement (Order in Council No. 1448-92) and its associated Manual of Implementation Guidelines (Ontario Ministry of Natural Resources, 1992). The policy has broad evaluation and protection goals to ensure no net loss of provincially significant wetlands. It also attempts to separate wetlands from the negative impacts of other land uses.

Wetland Policy 2.2 states that on adjacent lands, which are defined as all lands within 120 m of an individual wetland area and all lands connecting individual wetland areas within a wetland complex, "Development may be permitted only if it does not result in any of the following:

a) loss of wetland functions;
b) subsequent demand for future development which will negatively impact on existing wetland functions;
c) conflict with existing site-specific wetland management practices; and
d) loss of contiguous wetland area."

It is important to note that adjacent lands may extend more than 120 m from a wetland boundary within a wetland complex which may consist of two or more wetland areas, along with adjacent areas which are linked in a functional manner and are grouped within a common wetland boundary. These wetlands may be linked through large-scale hydrogeological processes which may extend in their influence far more than 120 m. Adjacent land by itself does not constitute a buffer zone, but identifies "an area within which no development may occur that will impair wetland functions." The actual impacts associated with a development on the criteria outlined in Policy 2.2 must be demonstrated by an Environmental Impact Study. It is through this process that effective buffer zones will be established. These buffers, however, are focused almost exclusively on some type and width of vegetative buffer designed to intercept contaminants, nutrients, and sediments carried by surface run-off from development sites. The use of buffers to intercept contaminants or attenuate temperature fluctuation in groundwaters are not indicated in the guidelines.

EXTRACTION ACTIVITIES AND IMPACTS

Aggregate in southern Ontario is extracted both from Quaternary sand and gravel deposits and from Palaeozoic bedrock formations. These materials have significantly different geotechnical characteristics, require different extraction technologies, and have different impacts on hydrogeological processes. Unconsolidated sand and gravel can be readily excavated with wheeled equipment and require little processing beyond sieving and washing. In contrast, bedrock formations must be extracted by blasting and processed at elaborate crushing and washing plants. Quarries are typically much larger and require much greater capital investment than sand and gravel pits which can be operated on a much smaller scale and with considerably less investment.

Extraction Activities

Both operation types are characterized by specific extraction activities ranging from the excavation and processing of material to maintenance and rehabilitation, which may have potential impacts on water resources. Some of these activities include:

A. **Site Preparation**
   Vegetation removal
   Watercourse diversion
   Overburden removal
   Site recontouring
   Berm construction

B. **Material Extraction**
   Excavation
   Blasting
   Dragline operation
   Water pumping from floor areas
   Pump discharge to surface waters

C. Material Processing
Pumping for wash operations
Storage and discharge of wash water
Secondary processing (concrete, asphalt)

D. Maintenance/Operations
Haul road treatment (oil, calcium chloride, water)
Vehicle and equipment fueling and maintenance
Material storage

E. Rehabilitation/After Use
Waste disposal
Lake creation
Wetland reestablishment
Intensive recreational (golf course) and residential uses

All of these activities may have direct or indirect impacts on water quality and quantity both at the surface and in the subsurface, and may impact the stability and function of wetland areas.

Water Quality Impacts Associated with Extraction

The quality of local surface and ground waters may be impacted through the introduction of contaminants, increases in turbidity from erosion and sedimentation, and increases in water temperature (Graves and De Vore, 1982). Specific extraction activities with greatest potential for these impacts can be summarized as follows:

Introduction of contaminants from:
Fuel and other material spills during operations,
Waste disposal,
Haul road dust suppressants,
Secondary processing sites,
Waterfowl fecal coliform inputs.

Increased sedimentation from:
Dragline extraction methods,
Increased erosion from stripped and extracted areas,
Water pumping to adjacent water courses,
Material washing and processing activities.

Increased temperatures from:
Increased water surface area and shallow water depth,
Decreased flow velocities and impoundments,
Decreased vegetation cover,
Decreased attenuation of seasonal fluctuations through ground water flow.

Water Quantity Impacts Associated with Extraction

The quantities of surface and ground water may also be affected by extraction as a result of the following activities:

Changes to ground water levels from:
Drawdowns related to pumping activities,
Seepage at perched water tables,
Increased local recharge in exposed areas,
Blasting impacts on aquifer and aquatard flow characteristics.

Changes to surface water volumes from:
Pump discharges to water courses,
Reduction of run-off and base flow contribution to water courses,
Increased capture of precipitation and local runoff in flooded areas,
Increased evaporation rates from the water surface of flooded areas.

Changes to ground and surface water flow patterns from:
Modification of local ground water hydraulic gradients,
Modification of local ground water flow rates,
Modification of material permeabilities,
Diversion and rechannelization of water courses.

An assessment of the impacts on wetlands of all these various activities requires a detailed understanding of the local water budget and the changes resulting from extractive activities. Figure 1 summarizes many of the impacts identified above and gives an overview of the complexities in the local water budget adjacent to extractive areas. These relationships must be understood in order to predict the possible impacts of extraction and to design appropriate mitigation and establishment of buffers.

Extraction Impact Scenarios

Extraction activities may have significantly different impacts on hydrogeological processes, depending on whether the lower limit of extraction lies above or below the level of the local ground water table. Based on the type of material being extracted and the relation of the extraction to the local water table elevation, four general extraction impact scenarios can be identified: sand and gravel may be excavated above or below local ground water levels, and similarly, bedrock extraction may take place above or below the water table. In all four scenarios, there may be impacts on the quality and quantity of both surface and ground waters.

By relating the four basic extraction scenarios with the two water-related impact areas, we can construct a matrix as shown in Table 1. This matrix identifies 16 possible impact situations and provides a qualitative overview of the relative

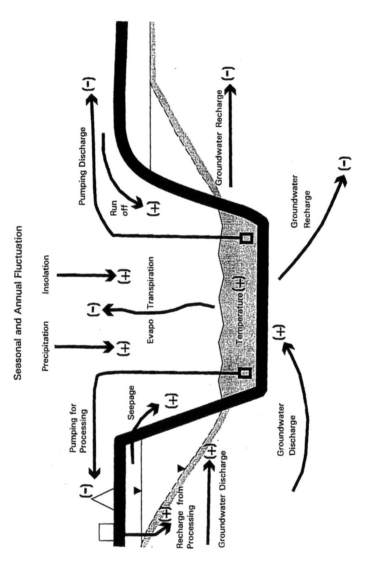

Figure 1   Schematic diagram of the possible water budget components for an extractive site.

Table 1 Aggregate Extraction Types and Impact Areas

Impact area extraction type	Water quality		Water quantity	
	Surface	Subsurface	Surface	Subsurface
Sand and gravel Above water table	Minor	Moderate	Minor	Moderate
Sand and gravel Below water table	Major	Major	Moderate	Major
Bedrock Above water table	Moderate	Moderate	Major	Moderate
Bedrock Below water table	Minor	Moderate	Major	Major

significance of these impact areas as minor, moderate, or major for the extraction scenarios. For each of the 16 situations, consideration of the extraction activities outlined above would produce a wide array of potential impacts. Some of these are summarized below.

Sand and Gravel Extraction Above Local Water Table

A number of Quaternary sand and gravel deposits such as eskers, kames, and morainic deposits are characterized by locally elevated topography so that extraction usually occurs above the local ground water table and the impacts of extraction on water quality and quantity are usually minor. The greatest potential for negative impacts on adjacent wetlands comes from the introduction of contaminants, water level changes from increased recharge associated with extraction, and the exposure of permeable materials by the extraction process.

Sand and Gravel Extraction Below Local Water Table

The most extensive and economically valuable sand and gravel deposits in southern Ontario are found in glacial meltwater channel or outwash deposits. These coarse and well-sorted aggregate materials are often occupied by the floodplains and adjoining fluvial terraces of modern watercourses and often support significant wetlands (Worthy *et al.*, 1985). Because of their location in discharge areas, pits often require extraction below local ground water levels. Consequently, impacts on water quality and quantity both at the surface and in ground water may be associated with this type of extraction. Examples of this situation include the Puslinch Township aggregate resource in the Mill Creek watershed (Gartner-Lee Ltd., 1989) and the Caledon Meltwater Channel gravels in Caledon and Erin Townships. This extraction scenario may pose the greatest risk for negative impacts on local wetland resources and is currently the focus for extensive monitoring programs being undertaken by extractive operators as part of license conditions.

Bedrock Extraction Above Local Water Table But Adjacent to Perched Overburden Water Tables

Many quarries require removal of considerable amounts of unconsolidated overburden, consisting of a variety of glacial materials, in order to expose bedrock for extraction. The overburden materials often contain local perched water tables which support wetlands. Gradual seepage from overburden into the quarry and the consequent lowering of the perched water levels over long periods of time may have significant impacts on wetlands. These impacts may be difficult to predict or measure over the short term but eventually may have some consequences for wetland function.

Bedrock Extraction in Local Unconfined Aquifers

A significant proportion of quarries are extracted below ground water levels, often in regionally significant unconfined aquifers and also at greater depths within confined aquifers. Extensive dewatering efforts are required to drain these quarries to permit extraction, with consequent local drawdown of ground water levels. Although this activity is often of great concern for the protection of drinking water supplies in regional aquifers, there may also be local impacts on wetlands which are hydrogeologically supported by these aquifers.

Other impacts associated with this type of extraction result from the discharge of pumped water to local surface watercourses and wetlands. Increased water levels and chemical changes which may be associated with different aquifer sources may have significant impacts on local wetlands (Moran and Cherry, 1981).

SHORT-TERM STRATEGIES FOR MITIGATION AND SEPARATION

In the short term, there are three site-specific phases of aggregate extraction where effective mitigation and separation strategies should be assessed: licensing and site planning, operations, and rehabilitation. More generally and over the longer term, regional resource management strategies which integrate aggregate extraction with other resource protection, management, and development objectives may have the greatest potential for the protection of significant wetland resources.

Considerable latitude exists in the Aggregate Resources Act (ARA) and the license application and site planning process for the consideration of impacts and the development of mitigation and separation strategies for the protection of wetlands. A range of site plans and conditions have been developed on a site-by-site basis over the past four years which provide a useful body of approaches.

Regulation of extractive operations is a major requirement of the ARA and is a responsibility of the Ministry of Natural Resources. Aggregate Resources officers in area offices are responsible for ensuring that extraction does not have negative impact on wetland functions. These staff work closely with area biologists and ecologists to ensure that wetland values are protected. There are also opportunities for innovative rehabilitation strategies to create high-quality wetland environments in extracted sites (Michalski *et al.*, 1987). There is a considerable body of literature from Britain and the United States on techniques for wetland creation in these circumstances (Svedarsky and Crawford, 1982).

LONGER TERM RESEARCH OPPORTUNITIES

In order to develop and assess mitigation and separation strategies between aggregate extraction sites and wetlands, site-specific experimental monitoring programs are required to quantify impacts of the four basic extraction scenarios. However, as a preliminary but important step, specific impacts could be studied through a review of monitoring programs which are already in place at currently licensed pits and quarries. A variety of monitoring programs have been initiated at extractive sites throughout southern Ontario as a result of MNR license review, Ontario Municipal Board recommendations, or at the initiative of licensees. Fruitful areas of investigation for retrospective and monitoring studies at these sites could include:

1. The impacts of overburden seepage from wetlands in perched water tables adjacent to quarries. Information may be used to establish appropriate buffer zones between extraction and perched wetlands.
2. The impacts of temperature increases in flooded areas associated with sand and gravel sites excavated below local water table levels. Information may be used to develop attenuation strategies and/or buffer zones to minimize water temperature increases in adjacent wetlands.
3. The impacts on adjacent wetlands of local increases in recharge and decreases in runoff associated with sand and gravel extraction above local ground water tables. Information may be used to develop site contouring and operations strategies to minimize negative impacts.
4. The impacts of quarry dewatering on local shallow unconfined aquifers which support wetland areas. Information may be used to develop pumping strategies to minimize water table lowering and also to minimize impacts associated with the discharge of pumped water back to local surface water courses and wetlands.
5. The impacts of sand and gravel extraction below the water table on local hydraulic gradients and flow patterns in both up and down gradient directions. Information may be used to develop strategies for the configuration of extracted areas to ensure that flow patterns and inputs to adjacent wetlands are not significantly altered.

It is also possible that some monitoring programs could be expanded in scope or modified to assess more general impacts related to extraction and wetlands, including:

1. The distinction between short-term and temporary impacts vs. long-term and permanent impacts.
2. The types and implications of cumulative impacts from a number of sites over a long period of time.

3. The identification of threshold levels for impacts.
4. Distinctions between local vs. regional impacts, on-site vs. off-site impacts, direct vs. indirect impacts, and immediate vs. delayed impacts.

Information derived from these monitoring programs may also be used to develop site plan review guidelines, guidelines for hydrogeological reports required by Section 9 of the Aggregate Resources Act, and criteria for determining when a Section 9 hydrogeological report should be prepared.

CONCLUSIONS

Over the longer term, strategic approaches are required for all of the ministry's resource management and protection mandates. A landscape ecology approach may provide a regional context for extraction while at the same time reducing the need for elaborate on-site mitigation as a result of more effective spatial separation on a landscape scale. The additional possibility of significantly increasing the extent of high-quality wetland areas through innovative approaches to the rehabilitation of extracted sites is also currently under review and holds the promise of creating significant amounts of wetland habitat. Aggregate extraction is an interim use of land. Careful planning to identify suitable extractive sites and innovative rehabilitation approaches to recreate natural environments may provide a valuable tool in the implementation of strategies to provide resource access while at the same time protect and enhance the natural heritage values of the settled landscapes of southern Ontario.

Because of the complex and poorly understood hydrogeological continuum which exists below the ground surface, the connections and relationships between extractive activities and nearby wetlands are difficult to predict or quantify. For these reasons the physical separation of the two at the surface through the use of specific buffer zones may not be a sufficient or useful approach to the protection of wetland functions. A review of mitigation strategies is required to identify and quantify successful approaches from the wide variety of approaches currently employed or in development by the extraction industry as part of the Ministry of Natural Resources' regulation activities. General criteria for mitigation can be identified, but effective strategies must be tailored to the unique circumstances of each operation.

More than 10% of currently licensed extractive sites in southern Ontario may be located in proximity to provincially significant wetlands; more than 30% in proximity to permanent watercourses; and more than one-third of all sites may have extracted or propose extraction below local ground water levels. Extraction activities in these situations have potential impacts on water quality and quantity, both at the surface and in ground water. Sand and gravel extraction below local water tables may have the greatest potential for negative impacts on adjacent wetlands, followed by bedrock extraction impacts on perched water tables and shallow unconfined aquifers. A range of mitigation and separation strategies are required to minimize wetland functions and areas affected by these activities.

REFERENCES

Gartner-Lee Ltd., 1989. Potential impact assessment consolidated report: ground water surface water aspects, proposed sand and gravel pit, Township of Puslinch, 210 p.

Geomatics International, 1993. Natural heritage system for the Oak Ridges Moraine area: GTA portion, Background Study No. 4 to the Oak Ridges Moraine Planning Study, 1993. Geomatics International, Burlington, Ontario, 78 p.

Government of Ontario, 1992. *Wetlands*. A statement of Ontario Government policy issued under the authority of Section 3 of the Planning Act, Lieutenant Governor in Council No. 1448/92, 1992.

Government of Ontario, 1990. *Aggregate Resources Act*, Revised Statutes of Ontario, 1990, Chapter A.8

Graves, D. H. and De Vore, R. W. eds., 1982. *Proceedings of the 1984 Symposium on Surface Mining, Hydrology, Sedimentology and Reclamation*, College of Engineering, University of Kentucky, Lexington.

Michalski, M. F. P., Gregory, D. R., and Usher, A. J., 1987. Rehabilitation of pits and quarries for wildlife, Ontario Ministry of Natural Resources, Toronto, 59 p.

Moran, S. R. and Cherry, J. A., 1981. The hydrologic response of aquifers at surface mine sites in western North Dakota, *Canadian Geotechnical Journal* 18:543–552.

Ontario Ministry of Natural Resources and Ministry of Municipal Affairs, 1992. *Manual of Implementation Guidelines for the Wetlands Policy Statement,* Toronto.

Ontario Ministry of Natural Resources, 1994. Mineral aggregates in Ontario, overview and statistical update 1992, Resource Stewardship and Development Branch, Toronto.

Ontario Ministry of Natural Resources, Hydrogeological guide for Aggregate Resources Officers, Resource Stewardship and Development Branch, unpublished report, Toronto.

Planning Initiatives Ltd. and Associates, 1993. Aggregate resources of Southern Ontario, a state of the resource study, Ontario Ministry of Natural Resources, Toronto.

Riley, J. L. and Mohr, P., 1994. The Natural Heritage of Southern Ontario's Settled Landscape. A review of conservation and restoration ecology for land-use and landscape planning, Ontario Ministry of Natural Resources, Southern Region, Aurora, Science and Technology Transfer, *Technical Report*, TR-001, 78p.

Svedarsky, W. and Crawford R. D., eds., 1982. Wildlife Values of Gravel Pits, Symposium Proceedings, University of Minnesota, Agricultural Experiment Station, Miscellaneous Publication 17, Minneapolis.

Worthy, W. E., Routly, D., and Harvey, E., 1985. Floodplain sand and gravel extraction in Cambridge District, Ontario Ministry of Natural Resources, unpublished report, Toronto.

17

THE CREAMS MODEL FOR EVALUATING THE EFFECTIVENESS OF BUFFER STRIPS IN REDUCING SEDIMENT LOADS TO WETLANDS

K. J. McKague, Y. Z. Cao, and D. E. Stephenson

ABSTRACT

Buffer strips are frequently suggested as a means to protect environmentally sensitive areas from nonpoint source loadings. At present, decisions on buffer widths are often based on minimum separation distances regulatory agencies specify between a sensitive area and the upland area generating the loadings. However, computer models do exist, and are able to simulate the pollutant-trapping efficiency of buffer strips. These models have the potential to assist decision makers in selecting optimum buffer widths for specific settings.

In this study, the CREAMS model was used to assess the effectiveness of buffers in protecting a wetland located downslope from an adjacent proposed urban development site. Data sets were prepared describing both existing site conditions and midconstruction conditions. Buffer strip widths were identified which the model predicted would bring sediment loading rates during the midconstruction periods down to levels matching the loading rates associated with the existing site conditions. The modeling exercise highlighted the importance of ensuring that runoff from the upstream area enters the buffer strip as uniform sheet flow to maximize buffer strip efficiency. The modeling also proved valuable in identifying points where sediment control techniques other than buffering were needed to fully protect the wetland from excess sediment loading as a consequence of urban construction activities.

1-56670-147-3/96/$0.00+$.50
© 1996 by CRC Press, Inc.

INTRODUCTION

Vegetated buffer strips have been suggested and applied in a variety of settings for the purpose of reducing nonpoint source pollution loadings to sensitive areas such as watercourses or wetlands. In agricultural settings for instance, grass filter strips have been installed downslope from croplands to control sediment, nutrient, and pesticide loadings to streams (Dillaha, 1989). Grass buffers have also been evaluated for their effectiveness in attenuating nutrient and pathogen loads associated with runoff from barnyards and manure stacks. Nonagricultural settings in which buffer strips or zones have been recommended for the purpose of protecting sensitive downstream areas include logging operations (Lammers-Helps and Robinson, 1991), construction sites, (Environment Canada, 1980), strip mining operations (Barfield *et al.*, 1979) and urban areas where there is a need to protect fish habitat from urban pollution sources (MNR, 1987).

Buffer strips can be defined as bands of planted or indigenous vegetation situated between pollutant source areas and receiving waters. Other names given to buffer strips include vegetated strips or vegetative filter strips, grass strips, buffer zones, grass buffer strips, and riparian plantings. Traditionally, buffer strips have been located immediately adjacent to watercourses. More recently, however, they have been suggested for use in a variety of other settings, such as along the lower boundary of agricultural fields or animal production facilities. By placing the buffer strips close to the pollutant source area, they provide localized erosion protection and filter out pollutants before runoff water begins to concentrate into flow pathways (Dillaha *et al.*, 1989).

The effectiveness of buffer strips in improving surface runoff water quality varies with many factors including the upstream catchment area, slope length and gradient above and within the buffer zone, volume and velocity of runoff water, the nature of the eroding sediment, vegetation type, height, and density, buffer strip width, and the characteristics of the transported pollutants. In many instances, only the width is taken into consideration when designing a buffer strip. In urban development settings for example, regulatory agencies often define a minimum set-back between a development area and a wetland or sensitive area. Depending on the physical and hydrologic characteristics of the upstream area, this minimum set-back requirement may be wider or narrower than is necessary to adequately protect the downstream area. The nonpoint source pollution models can take into consideration a range of different factors in arriving at an optimum buffer width.

BUFFER STRIP MODELING AND APPLICATION OF THE CREAMS MODEL

Many researchers have undertaken investigations to quantify the effectiveness of grass buffer strips. As a result, several models have been developed to assist with vegetative filter strip design and evaluation. GRASSF, for example, is an event-based model developed for designing vegetative filter strips for sediment removal applications (Barfield *et al.*, 1979; Hayes *et al.*, 1979). This model was developed and tested by using runoff plot data from multiple rainfall events. While in model development tests, sediment removal predictions of the GRASSF model were shown to be very similar to observed values, one limitation to the model is that it does not consider sediment deposition from the water ponded upslope of the grass strip. Research has shown this upslope ponding area to be the zone of greatest deposition associated with filter strips (Hayes and Hairston, 1983). By overlooking the deposition of sediment in the upslope ponded water, the GRASSF model tends to underpredict the trapping capability of buffer strips.

The GRAPH model, developed by Lee *et al.* (1989), overcomes some of the limitations inherent in the GRASSF model. It also simulates nutrient as well as sediment transport in vegetative filter strips. The GRAPH model takes into consideration the effects of nutrient advection (transport by water flow) as well as adsorption and desorption processes on nutrient transport. The model also considers the effect of different vegetative cover types on filter strip effectiveness.

The CREAMS model is a more comprehensive model than the forementioned models. CREAMS was developed by a task force formed by the Agricultural Research Service (ARS), a branch of the United States Department of Agriculture (USDA), to evaluate nonpoint source pollution from watersheds. The CREAMS model is a physically based, daily simulation model that estimates runoff, erosion/ sediment transport, as well as plant nutrient and pesticide losses from field-size areas. A complete description of model use and the theory behind its algorithms, including those used to simulate buffer strip effectiveness, is provided by Knisel and Nicks (1980). A general overview of the model's operation is presented in Figure 1.

Other researchers have reported good success using the CREAMS model for evaluating buffer strip effectiveness. Flanagan *et al.* (1989) used CREAMS to examine the variability in buffer strip effectiveness as affected by different parameters and verified the model-generated output with field-measured data. Williams and Nicks (1988) used the CREAMS model to evaluate the trapping of sediment by grass filter strips along overland and concentrated flow pathways and found it to be a useful prediction tool.

This study applied the CREAMS model to estimate the buffer width required to protect a portion of the Eastview wetland located in Guelph, Ontario from proposed adjacent urban development activities. Concern at the Eastview site centered around the potential for a significant rise in sediment loadings to the wetland during the critical midconstruction phase of the development of this site. The midconstruction phase is the period of time where typically the vegetation and topsoil from the development area is stripped, making the site particularly vulnerable to erosion. The CREAMS model was utilized to estimate the minimum buffer width required at various points around the wetland to ensure that sediment loadings did not exceed existing levels.

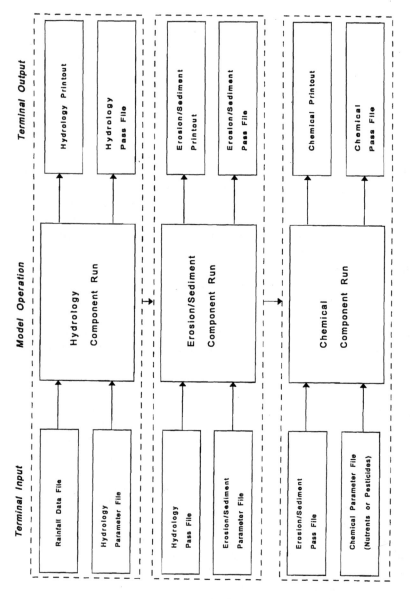

Figure 1   Flow chart of CREAMS model operation.

THE EASTVIEW WETLAND EXAMPLE

The Setting

A plan view, illustrating the topography of the proposed development area and the location of the sensitive wetland area to be protected, is provided in Figure 2. Following a site visit and a review of the proposed construction plans for the area, vulnerable sections of the wetland were identified and subwatersheds draining to these vulnerable areas from the proposed development site were delineated. The subwatersheds of concern, as shown in Figure 2, were assigned a letter code for identification purposes.

The four subwatersheds identified, each draining its own portion of the proposed urban development site, are unique with respect to their topographic and hydrologic features. Watershed A may be characterized as having moderate and relatively short slopes which result in runoff water traveling in a well-distributed and uniform pattern from the watershed to the wetland. Watershed B has moderate to steep slopes with the runoff water concentrating in a defined overland flow pathway prior to entering the wetland. Land slope in Watershed C is quite gradual and, like Watershed A, runoff from the watershed enters the wetland as uniform sheet flow. Finally, Watershed D can be characterized as having long moderate slopes with the runoff water again being directed as sheet flow towards Hadati Creek. A more detailed summary of the parameters describing the four watersheds modeled in this study is provided in Table 1. Flow paths for each watershed are shown in Figure 2.

Data Set and Procedure

Daily precipitation records for a 10-year period, 1979 to 1988 inclusive, were obtained from a local weather station and used as input to CREAMS' hydrology component. The CREAMS model was used to predict sediment loading rates for each of the following:

- Existing site conditions
- Midconstruction site conditions without buffers
- Midconstruction site conditions with 5', 10', 25', 50', 100', and 200' buffers placed between the development area and the wetland

The input parameters which were modified in order to develop data sets representing these various conditions included entries for the Universal Soil Loss Equation (USLE) cover (C), soil erodibility (K), and contour (P) factors, Manning's n, soil infiltration rates, the relative proportion of overland flow and channel flow pathway lengths receiving the buffer treatment, and the proportion of bare subsoil still exposed within the watershed following buffer strip installation.

The preconstruction cover was defined as continuous corn crop. For midconstruction simulations, a bare soil situation was assumed. Topsoil was also

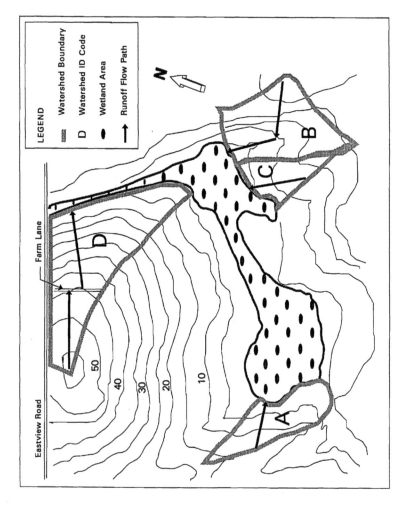

Figure 2 Plan view of modeled site.

Table 1   Some Parameters of the Four Watersheds

	Watershed A	Watershed B	Watershed C	Watershed D
Area (acres)	6.9	12.0	4.6	18.2
Average slope (%)	4.0	9.3	5.3	5.8
Slope (%)	3.0–11.3	3.1–12.0[a]	5.3–5.3	3.1–10.8
Slope length (ft)	377	703	492	1,171
Soils	London Loam	Parkhill Loam	Parkhill Loam	London Loam
Slope/runoff profile	Overland	Overland-channel	Overland	Overland

[a] The 3.1% slope represents the channel profile's slope gradient.

assumed to have been stripped from the area by the midconstruction stage of site development. Thus soil characteristics used as input for the midconstruction simulations were those associated with what would normally be the existing soil's C horizon. In some cases, watershed size and runoff slope lengths and gradients changed between the existing and the midconstruction simulations. For example, Watershed D had a farm laneway across its width which acted as the upper boundary for the watershed under existing site conditions (see Figure 2). For the midconstruction situation, however, it was assumed that this farm laneway would be removed, thus increasing slope length and watershed area. The CREAMS model runs simulated sediment loading to the wetland over a 10-year period, and from these values an average annual sediment loading value to the wetland from the watershed was estimated.

Results and Discussion

The model predictions of existing and midconstruction sediment loads from each watershed were compared to assess the degree to which sediment loading to the wetland would increase if the site underwent urban development. The difference between the existing conditions and midconstruction sediment loads also provided an indication of how effective a buffer strip in the area should be in order that midconstruction conditions match existing sediment loading levels. Figure 3 graphically shows the difference between existing and midconstruction annual sediment loading rates to the wetland from Watershed D. These results suggest that the buffer to be installed in Watershed D should be capable of reducing the sediment loading rate by 23 tons/acre/year during the construction stage of site development so as to maintain predevelopment conditions. Note that such high sediment loading rates would not be expected to continue indefinitely because once construction activity ceased, sediment loading levels would drop significantly.

Figure 4 summarizes the predictions of sediment loading for each of the watersheds and for each of the buffer strip widths simulated. In three of the four watersheds modeled, the sediment loading rates were predicted to decline significantly following installation of a buffer strip. The model results suggested that the first 10 to 20 ft of buffer strip width were particularly beneficial. As buffer widths increased, there was a corresponding gradual decline in the incremental gains made in sediment-trapping benefits. The exception to this general trend among the

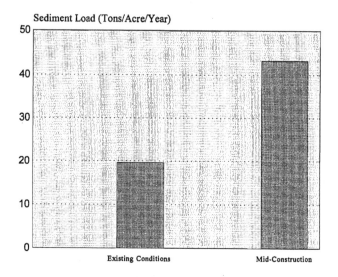

Figure 3   Sediment loadings to Eastview Wetland from Watershed D.

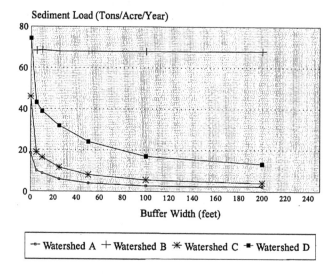

Figure 4   Buffer strip effectiveness. Watershed A: moderate slope, overland flow; Watershed B: short gradual slope, overland flow; Watershed C: moderate slope, channel flow; Watershed D: long moderate slope, overland flow.

watersheds was associated with Watershed B. There was essentially no attenuation of sediment loadings over the entire range of buffer widths considered on this watershed. This is due to the fact that runoff water exits Watershed B as a concentrated flow rather than as sheet flow.

The minimum buffer width necessary to reduce sediment loadings from the construction stage of site development to existing loading levels can be quickly

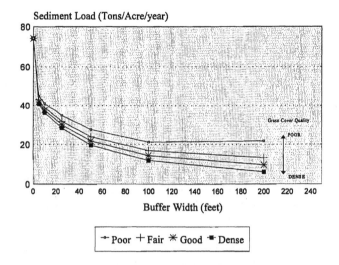

Figure 5 Impact of grass quality on buffer strip effectiveness — Watershed D.

determined from the CREAMS model estimates presented in Figure 4. For example, to reduce loading rates from Watershed D to the existing rate of 19.6 tons/acre/year (see Figure 3), the CREAMS model would suggest the installation of a buffer at least 90 ft wide (see Figure 4). Similarly, the CREAMS model indicated that installing a buffer in Watershed B would be ineffective, regardless of its width, and that other sediment control measures were needed to reduce sediment loadings to the wetland from Watershed B.

The CREAMS model was also used to predict the effect of grass quality on buffer strip performance. Four different general grass cover conditions within the buffer zone were modeled, ranging from poor grass cover to dense cover. The CREAMS model allows the user to enter a Manning's n value which best represents the quality of the grass or vegetative cover condition being simulated. Figure 5 illustrates the effect of altering the Manning's n value on model output and points out the need to consider the buffer strip's vegetative quality when assessing the effectiveness of an installed buffer strip in the field. Similarly, when using the CREAMS model to predict the effectiveness of a buffer, it is important that the Manning's n value which best represents the buffer's vegetative condition be selected. Dense strips of grass provide a much higher degree of flow retardance compared to poor stands, making the dense stands more effective at filtering out pollutants. Recently established buffer strips that have not had time to develop a dense sod cover, will not be as effective as older well-maintained buffers.

SUMMARY AND CONCLUSIONS

Generally, an important criterion for designing buffer strips is the minimum separation distance between an environmentally sensitive area and the development area. Little consideration is given to the other factors affecting buffer strip sediment trapping efficiency. As a consequence, the buffer strips installed may be actually overdesigned or underdesigned from a sediment-trapping perspective. There is merit in using existing nonpoint source models which can help in the consideration of more factors that will influence buffer strip effectiveness. The important findings from the modeling exercise conducted on the Eastview wetland development site were as follows:

1. While models have their limitations, use of the CREAMS model did provide a scientifically based approach for selecting a buffer strip width which could achieve the desired reduction in sediment loadings to a sensitive area.
2. Buffer strips are most effective at reducing sediment loadings from runoff originating from upland areas. As the runoff enters the buffer zone as dispersed sheet flow, the bulk of the sediment drop occurs in the first few feet of buffer width.
3. Buffer strips are ineffective at reducing sediment loadings from surface runoff moving in a concentrated or channelized form prior to entering the buffer strip. In these situations, alternate erosion control techniques need to be implemented to reduce nonpoint source pollutant loadings to the sensitive area being protected.
4. Modeling can help identify points where buffer strips are relatively ineffective as a sediment control technique. This is particularly useful in urban development settings where both land area available for development and sensitive environment areas are highly valued.

Buffers which are sized using a soil erosion and sedimentation simulation model such as CREAMS consider only sediment and selected pollutant loading impacts on sensitive areas. They do not consider factors concerned with maintaining or enhancing wildlife habitat. Consequently, applying the results from a CREAMS simulation is not applicable in all situations. In circumstances where a more detailed scientifically based assessment is needed to evaluate the relative benefits of alternative buffer widths on pollution attenuation, such a modeling exercise can be quite appropriate. The potential also exists to use output from a series of theoretical situations for the purpose of developing design charts, eliminating the need for all buffer strip designers to be familiar with the operation and use of models such as CREAMS. This information could in turn be combined with other factors, including wildlife habitat requirements, to assist in selecting the optimum buffer width addressing all points of concern.

REFERENCES

Barfield, B. J., Toollner, E. W., and Hayes, J. C., 1979. Filtration of sediment by simulated vegetation. I. Steady-state flow with homogenous sediment. *Transactions of the American Society of Agricultural Engineers.* 22:540–548.

Dillaha, T. A., Sherrard, J. H., Lee, D., Mostaghimi, S., and Shanholtz, V. O., 1985. Sediment and phosphorus transport through vegetative filter strips: Phase I, Field Studies. *American Society of Agricultural Engineers.* Paper No. 85-2043.

Dillaha, T. A., Sherrard, J. H., Lee, D., Mostaghimi, S., and Shanholtz, V. O., 1988. Evaluation of vegetative filter strips as a best management practice for feed lots. *Journal of the Water Pollution Control Federation.* 60:1231–1238.

Dillaha, T. A., Reneau, R., Mostaghimi, S., and Lee, D., 1989. Vegetative filter strips for agricultural non-point source pollution control. *Transactions of the American Society of Agricultural Engineers.* 32:513–519.

Dillaha, T. A., 1989. Water quality impacts of vegetative filter strips. *American Society of Agricultural Engineers.* Paper No. 89-2043.

Environment Canada, 1980. Environmental Code of Good Practice for General Construction. Report No. EPS 1-EC-80-1

Flanagan, D. C., Neibling, W. H., and Burt, J. P., 1989. Simplified equations for filter strip design, *Transactions of the American Society of Agricultural Engineers.* 32:2001–2007.

Hayes, J. C., Barfield, B. J., and Barnhisel, R. I., 1979. Filtration of sediment by simulated vegetation. II. Unsteady flow with non-homogenous sediment. *Transactions of the American Society of Agricultural Engineers.* 22:1063–1067.

Hayes, J. C. and Hairston, J. E., 1983. Modelling the long-term effectiveness of vegetative filters on on-site sediment controls. *American Society of Agricultural Engineers.* Paper No. 83-2081.

Kao, D., Barfield, B. J., and Lyons, A. E., 1975. On-site sediment filtration using grass strips. *National Symposium on Urban Hydrology and Sediment Control.* University of Kentucky, Lexington, pp. 74–82.

Knisel, W. G. and Nicks, A. D., 1980. CREAMS: a field scale model for chemicals, runoff and erosion from agricultural management systems. Volume I, model documentation. U.S. Department of Agriculture, Washington, D.C.

Lammers-Helps, H. and Robinson, D. M., 1991. Literature review and summary pertaining to buffer strips. Final report. Research Sub-program of the National Soil Conservation Program, Agriculture Canada, Guelph, Ontario.

Lee, D., Dillaha, T. A., and Sherrard, J., 1989. Modelling phosphorus transport in grass buffer strips. *Journal of Environmental Engineering.* 115:409–427.

Ministry of Natural Resources (MNR), 1987. Guidelines on the use of "vegetative buffer zones" to protect fish habitat in an urban environment. Ontario Ministry of Natural Resources, Toronto.

National Research Council, 1993. *Committee on Long-Range Soil and Water Conservation Board on Agriculture, Soil and Water Quality: An Agenda for Agriculture.* National Academy Press, Washington, D.C.

Stearns, L. J., Ackerman, E. O., and Taylor, A. G., 1982. Illinois vegetative filter design criteria. *American Society of Agricultural Engineers.* Paper No. 82-2613.

United States Department of Agriculture, Soil Conservation Service. 1984. User's guide for the CREAMS computer model — Washington computer center version, USDA, Washington D.C.

Williams, R. D. and Nicks, A., 1988. Using CREAMS to simulate filter strip effectiveness in erosion control, *Journal of Soil and Water Conservation.* 43:108–112.

18

THE USE OF VEGETATIVE BUFFER STRIPS TO PROTECT WETLANDS IN SOUTHERN ONTARIO

A. J. Norman

ABSTRACT

Approximately 70% of wetlands present in southern Ontario at the time of settlement have been lost to human uses. Significant development pressures continue on the remaining wetlands. The use of vegetated buffer strips of various widths has been recommended in the literature to protect wetland resources from adjacent land uses. These vegetated areas retain some of the pollutants and thereby reduce degradation of water quality and wildlife habitat resulting from adjacent land uses. The Ontario Ministry of Natural Resources and other regulatory agencies require guidance on the width of vegetated buffer strips to help reduce the negative impacts of developments. To help protect the remaining wetlands, recommendations are made on various widths of buffer strips based on a review of the literature.

INTRODUCTION

Major losses of wetlands have occurred in southern Ontario from the time of settlement. By 1982, 68% of wetlands in southern Ontario south of the Precambrian shield had been converted to other land uses (Snell, 1987). Pressure on these important natural features and adjacent lands in southern Ontario continues to be intense.

Vegetated buffer strips are one of the mechanisms used to protect wetlands and other aquatic resources from the negative effects of adjacent land uses. Buffer strips are strips of vegetated land composed in many cases of natural ecotonal and upland plant communities which separate development from environmentally sensitive areas and lessen these adverse impacts of human disturbance (Shisler *et*

1-56670-147-3/96/$0.00+$.50
© 1996 by CRC Press, Inc.

al., 1987). A literature review has been completed on the use of vegetated buffer strips in the protection of aquatic resource features from human developments and activities. Recommendations on the use of vegetated buffer strips to protect wetland functions and values are made.

FUNCTIONS OF BUFFER STRIPS

Buffer strips adjacent to wetlands serve to protect a wide range of wetland functions and values from the negative impacts of developments and human activities. The protection afforded by the buffers could be derived from a physical, chemical, or biological process or a combination of these processes. The following sections deal with buffers in relation to water quality and wildlife.

Water Quality

Development activities expose soil to erosion during construction, and subsequent human uses of the area continue the negative effects of soil erosion and sedimentation. Buffer strips have been used and are recommended for the protection of aquatic resources such as lakes, rivers, streams, and wetlands from detrimental effects of forestry, agriculture, and urbanization.

Forestry

Concern has been expressed in the forestry literature for over 40 years on the effects of logging practices on erosion and sedimentation. Croft and Hoover (1951), working in the Rocky Mountains, observed that even the most careful logging operations greatly disturbed the soil. One of the "common sense" improvements they recommended to lessen damage to permanent streams by logging operations was to log lightly in a streamside strip 62 to 123 m wide. Recommended buffer widths from selected forestry studies are summarized in Table 1.

Trimble and Sartz (1957) measured the distance sediment moved from a logging road in a well-stocked hardwood forest in the White Mountains of New Hampshire. For the situation where maintenance of the highest possible water quality is of paramount importance, such as in municipal watersheds, they recommended starting with a strip 15.2 m wide on level land and increase it by 0.6 m for each 1% increase in slope.

Packer (1967) studied criteria for design, location, and construction of logging roads in the Rocky Mountains to prevent damage to the water resource and to conserve soil. The author determined that a buffer strip of 39 m of herbaceous cover would contain 83.5% of sediment flows from roads on sites with the most stable soil.

For protection of small streams from logging practices in the Maritimes, van Groenewoud (1977) concluded that a 15-m buffer strip on relatively flat terrain would provide adequate protection against sediments. However, in New Brunswick, mud generated from road construction in an area with springs, flowed across a

TABLE 1 Buffer Widths Reported from Selected Forestry Studies and Review Papers

Source	Location	Buffer width (m)	Slope (%)	Comments
Croft and Hoover (1951)	Colorado, Utah	61.5–123.1	—	Recommend careful logging practices
Trimble and Sartz (1957)	New Hampshire	15.2	0	
		21.2	10	
		27.7	20	
		33.2	30	
		39.2	40	
		45.2	50	
		51.2	60	
		57.2	70	
Packer (1967)	Utah	39.1	—	Herbaceous buffer strip of this width would remove 83.5% of sediment
van Groenewoud (1977)	New Brunswick	15–65	—	From literature review and local research
Corbett et al., 1978	Eastern United States	12.3–30.8	—	Review paper
Clinnick (1985)	Australia	30 average (6–88 m)	—	Review paper

100-m wide buffer strip with a 3% slope. The author concluded that if areas with springs are protected, a 65-m buffer strip should be sufficient on most slopes. However, it was cautioned that each case would have to be judged separately. As well, the buffer strip should extend the full length of the stream and also protect the major springs.

Corbett et al. (1978), reviewed the literature pertaining to timber harvesting practices and water quality in the eastern United States. Strip widths varied with conditions on different watersheds. Most common widths were from 12.3 to 30.8 m on each side of the stream. Even though, a strip 20.3 m to 30.8 m was usually needed to protect the stream ecosystem. It was noted that a wider streamside management zone may be needed where slope or soil conditions dictate.

Clinnick (1985) reviewed literature from Australia, North America, and New Zealand on the use of buffer strips to protect streams and maintain water quality as a result of forest management practices. Recommendations for buffer strip widths ranged from 6 to 88 m. The most commonly recommended width for stream buffers was 30 m. This width should increase with increasing site limitations. Recommendations on buffer width and extent are best made by taking into account all of those factors pertaining to the particular site. It was noted that even within the same region, variations in climate, soil, physiography, and type of forest operations necessitate differing site-specific recommendations for catchment protection.

Barker (1983) observed that little research had been completed to test the effectiveness of buffer zones in protecting forest streams from sedimentation generated from logging practices throughout the forest. Barker concluded that it had not yet been demonstrated that a streamside buffer zone of any configuration is an effective means of precluding sediment delivery to forest streams. Sheet flow of runoff water does not predominate in the forest system. Water is concentrated in the rills and gullies that make up the ephemeral drainage system of the basins, and buffer zones should be located in these areas. Their main value is in their ability to reduce rather than to preclude the delivery of sediment to the stream.

Several authors have warned about the need to practice careful logging practices in addition to using buffer strips to control erosion and sedimentation (Croft and Hoover, 1951; Horneck and Reinhart, 1964; Haupt and Kidd, 1965; van Groenewoud, 1977; Barker, 1983). Dissmeyer and Singer (1977) observed that filter strips are effective only if the volume of erosion and runoff from above does not exceed the capacity of the filter to absorb it. They concluded that erosion and runoff must be controlled at their sources and recommended that the area above the filter should have a prescribed percentage of ground cover depending on the slope and fragility and stability of soils.

Agriculture

The majority of the research efforts have been focused on sediment and sediment-bound nutrients in herbaceous buffer strips. Relatively little work has been done on soluble nutrients, pesticides, or pathogens. The majority of the experiments have involved the use of rainfall simulation on small source plots adjacent to grass buffer strips (Lammers-Helps and Robinson, 1991).

Wilson (1967) examined the use of buffer strips to remove pollutants from water entering reservoirs and found that 3 m was sufficient to remove the maximum percent of sand, 15.2 m for silt, and 122 m for clay. The data showed an inverse relationship between the width of the filtration strip and the diameter of the particle. Colloid-size materials require a wider filter strip for maximum deposition than silt. The author concluded that grass filtration is an effective, economical, first-stage procedure for reducing sediment in flood water. The experiments showed that width of strip, initial turbidity, application rate, slope, grass height, degree of ramification of vegetation, and degree of submergence are interrelated factors affecting sediment removal. Buffer widths and level of contaminant removal from various agricultural studies are summarized in Table 2.

Paterson et al. (1980) studied the management of dairy liquid waste from a dairy operation in Illinois. A 71% reduction in total suspended solids in a 35-m fescue filter strip was observed and the most effective way to minimize these contaminants was to increase filter width. Young et al. (1980) tested the effectiveness of vegetated buffer strips to control pollution from feedlot runoff. Buffer strips of 36 m reduced to acceptable levels concentrations of both nutrients and microorganisms in feedlot runoff.

Dickey and Vanderholm (1981) evaluated four vegetative filters on feedlots in Illinois. About 70% reduction in concentration of contaminants occurred after flowing through a 90-m-wide vegetative buffer strip. However, discharge concentrations did not meet stream quality standards. The buffer width required to meet standards was postulated to be two to four times wider than those evaluated.

Dillaha et al. (1985) evaluated vegetative filter strips for sediment and phosphorus removal from surface runoff arising from feedlots. Sediment removal

Table 2 Buffer Widths from Various Agricultural Studies

Author (date)	Location	Nature of experiment	Slope (%)	Distances (m)	Comments
Wilson (1967)	Arizona	Remove sediment from water entering reservoirs	1	3–122	Sand-largest % of sediment at 3 m Silt-largest % of sediment at 15.2 m Clay-largest % of sediment at 122 m
Young et al., (1980)	Minnesota	Pollution from feedlot runoff	4	36	2 Years of testing
Paterson et al., (1980)	Illinois	Pollution from diary operation	—	35	71% Reduction in total suspended solids
Dickey and Vanderholm (1981)	Illinois	Pollution from feedlots	2	90	70% Reduction in concentration of pollutants
Dillaha et al., (1986a)	Virginia	Sediment and phosphorus from feedlot runoff	—	9.1	91% Reduction in solids 69% Reduction in phosphorus
Dillaha et al., (1986b)	Virginia	Survey of 33 Virginia farms			36% Of filter strips totally ineffective

rates were high during the first few runs but decreased during subsequent runs. Concentrated flow across the filters dramatically decreased filter performance. Vegetative filters were less effective for phosphorus removal than for sediment.

A 9.1-m filter strip with shallow uniform flow removed 91% of incoming solids and 69% of incoming phosphorus from feedlot runoff. Vegetative filters with concentrated flow were much less effective than shallow uniform flow plots (Dillaha *et al.*, 1986a).

Of the vegetative filter strips on 33 Virginia farms, 36% were found to be totally ineffective in controlling pollution. They were effective only if flow is shallow and the filter strip is not submerged. Grasses and legumes are most effective for erosion control and water quality improvement because of their dense growth, resistance to overland flow, and filtering ability. Dillaha *et al.* (1986b) recommended removing trees, stumps, brush, and similar materials to avoid interference with proper filter strip operation and maintenance. They observed that filter strips useful as wildlife habitat are totally inadequate for effective filtering or flow retardance as vegetation is too sparse.

Magette *et al.* (1986) studied the use of vegetated filter strips in removing nutrients and sediment from agricultural runoff. The performance of filter strips in reducing nutrient losses was highly variable. The strips were less effective as time went on in reducing nutrient losses. The authors conclude that the performance of vegetated filter strips is probably much less than expected in actual use.

Dillaha (1989), in a review paper on the removal of sediment and nutrients from cropland runoff by vegetative filter strips, has observed that they are effective only if flow is shallow and uniform and the strips were not previously inundated with sediment. Effectiveness decreases with time as sediment accumulates. Soluble nutrients are not removed as effectively as sediment. Most research on farm-vegetated filter strips has been of a short-term nature and did not consider the long-term effectiveness for pollutant reduction, and therefore these studies will overpredict the effectiveness of vegetated filter strips over the long run.

Urbanization

A filter strip to work effectively in an urban environment should (1) be equipped with some sort of spreading device, (2) be densely vegetated with a mix of erosion-resistant plant species that effectively bind the soil, (3) be graded to a uniform, even, and relatively low slope, and (4) be at least as long as the contributing runoff area (Schueler, 1987).

The rate of removal of pollutants appears to be a function of the width, slope, and soil permeability of the strip, the size of the contributing runoff area, and the runoff velocity. From a design standpoint, the only variables that can be effectively manipulated are the width and slope of the strip. Strips ranging in size from 30.5 to 91.5 m are probably needed for adequate removal of the smaller-size sediment particles found in urban runoff (OWML, 1983).

Filter strips will not function as intended on slopes greater than 15%. These steeper slopes should still be vegetated but off-site runoff should be diverted around rather than through them. Filter strip performance is best on slopes with a grade of 5% or less (Schueler, 1987).

All filter strips should be inspected on an annual basis. Strips should be examined for damage by foot or vehicular traffic, encroachment, gully erosion, density of vegetation, and evidence of concentrated flows through or around the strip. Filter strips are relatively inexpensive to establish, particularly if preserved before the site is developed (Schueler, 1987).

Wildlife Habitat

Vegetated buffer zones at the edge of wetlands have characteristics of riparian zones (Thomas *et al.*, 1979; Oakley *et al.*, 1985). Riparian zones comprise a minor portion of the overall area and are generally more productive in terms of biomass than surrounding upland areas. They are a critical source of diversity within the ecosystem and they provide considerable amounts of food, water, and shelter.

Vertical layers of herbaceous vegetation, shrubs, and trees often occur in a relatively small area adjacent to wetlands. This stair-stepping of contrasting forms of vegetation provide diverse nesting and feeding opportunities for wildlife — especially birds and bats. The association of particular birds with distinct layers of vegetation has been well demonstrated (MacArthur *et al.*, 1962). In general, wildlife uses riparian zones disproportionately more than other types of habitat (Kauffman, 1988; Oakley *et al.*, 1985; Thomas *et al.*, 1979).

Mammals

Compton *et al.* (1988), working in Montana, concluded that the amount of riparian cover may determine the potential number of white-tailed deer that bottom land habitats can support. Any substantial decrease in riparian cover through interruption of succession of riparian communities as a result of land clearing or logging of riparian timber may reduce the potential for sustaining deer (Compton *et al.*, 1988). In southern Ontario, lowland swamps can be valuable deer wintering areas (Voigt, 1990).

Dickson (1988) investigated the effects of the extent and composition of streamside zones on vertebrate wildlife communities in pine forests in the southern U.S. The animal populations were related to the width of streamside zones. Virtually no squirrels or squirrel nests were found in streamside zones less than 40 m wide, but were common in those zones >50 m wide. Streamside zones narrower than about 50 m do not appear capable of supporting permanent resident populations of squirrels.

Birds

Forty species of birds are dependent on wetlands in southern Ontario (Glooschenko and Grondin, 1988). Waterfowl, particularly mallards, blue-winged teal, wood ducks, and Canada geese do well in southern Ontario. In 1986, southern Ontario provided habitat for over 79,000 mallard and 12,300 blue-winged teal successful breeding pairs. This represented 29% of the Ontario total for the species (Glooschenko and Grondin, 1988).

Large blocks of vegetation adjacent to marshes are important for breeding waterfowl. Dabbling ducks nest on the ground, concealed by long vegetation. Large areas of cover are required by females to escape detection by predators. Narrow strips of vegetation act as predator traps, concentrating nesting females and predators into the same small area. Consequently, increased predation operates to decrease production of young ducklings. If predation is severe enough, along with other mortality factors, recruitment is not high enough to sustain the population.

Historically, or at least when land use pressures were less intense, it seems that nesting success was relatively high. Kalmbach (1939) found the average nesting success to be 60% in 22 areas across North America. Anderson (1957) and Keith (1961) found nesting success in two separate studies to be 38.4 and 39%, respectively. Cowardin et al. (1982) state that nest success estimates for the northern U.S. are in the range of 5 to 15%.

Changes in habitat and predator populations seem to be responsible for decreasing nesting success. Increasingly intensified agriculture has decreased both the availability of nesting cover for dabbling ducks and the habitats available to resident prey. Consequently, nesting ducks and foraging predators have been forced into the remaining untilled grassland (Higgins, 1977; Cowardin et al., 1982).

Duebbert (1969), Jarvis and Harris (1971), and Nelson and Duebbert (1973) demonstrated that large blocks of dense residual herbaceous cover close to wetlands attracted higher densities of nesting waterfowl. Significantly higher nesting success was found for ducks using undisturbed grass-legume cover on retired fields compared to land modified by arable agriculture. Fields 12 to 54 ha in size of cool-season-introduced grasses and legumes provided relatively secure nesting cover for dabbling ducks without predator control. For optimum habitat, established cover should remain completely undisturbed for 5 to 10 years (Duebbert and Lokemoen, 1976). Duebbert and Kantrud (1974) recommend retention of large blocks of nesting habitat in which ducks can disperse to nest as an effective means of reducing nest loss to predators.

Other recent work has confirmed the detrimental effects of loss of large areas of nesting habitat to nesting ducks and the need to manage for this type of habitat to retain waterfowl populations (Blohm et al., 1987; Greenwood et al., 1987; Sparrow and Patterson, 1987; Hochbaum et al., 1987; Turner et al., 1987).

Most of the published information on nesting of dabbling ducks is from the prairies. However, the two dabbling species (mallard and blue-winged teal) most dependent on upland nesting area adjacent to southern Ontario marshes likely respond very similarly to habitat cues in Ontario as they do on the prairies. Both species migrate, with pairing of birds from other parts of North America occurring on the wintering grounds and on the flyways. A regular mixing of genes, behavior and individuals from one part of the country to the other occurs. Consequently, a limited nesting habitat in Ontario is likely to affect these two species in a manner similar to those on the prairies.

Distances from nests found to marsh edges are an important consideration in establishing vegetative buffer strips of adequate width as nesting cover. Several studies have shown that the mean distance from mallard nests to the edge of the adjacent wetland was over 200 m (Labisky, 1957; Livezey, 1981; Smith, 1953). In general, blue-winged teal nest closer to the wetland edge than mallards.

It is reasonable to assume that for each of these studies, the mean distance of 200 m included approximately half of the nests. The other half of the nests occurred beyond this mean distance. Therefore, to provide for most mallards a block of nesting cover extending well beyond 200 m from the edge of the wetland would need to be provided. A distance of 300 m should include most nests, providing a block of nesting cover large enough to make detection by predators difficult.

Limited quantitative data are available for some other bird species. Stauffer and Best (1980) studied avifauna of riparian communities during late spring and early summer in Iowa. Bird species richness increased with the width of wooded riparian habitats. Of 17 species observed that had earlier been reported to be limited by habitat patch size, 7 were found breeding in relatively narrow (≤20 m wide) habitat patches; 11 were found breeding in habitat patches ≤40 m wide; 9 other species were found breeding in relatively wide patches of riparian habitat (90 to 200 m).

Kellar *et al.* (1993) surveyed birds in riparian forests in Maryland and Delaware to assess if presence of any species was dependent on corridor width. Corridors less than 50 m wide did provide habitat for many edge species. Ten species, including several neotropical migrants, showed a statistically significant increase in probability of occurrence as corridor width increased. The authors recommended that riparian forest be at least 100 m wide to provide some nesting habitat for area-sensitive species.

In forested areas the Ontario Ministry of Natural Resources has recommended that a minimum of 50 m on either side of a river or lake be left uncut. This consideration is recommended for accipiters, buteos and eagles, cavity nesting birds, and warblers (James, 1984a, 1984b, 1984c; Allan and Bohart, 1979; Evans and Connor, 1979).

Amphibians and Reptiles

Some 28 species of amphibians and reptiles rely on wetlands in southern Ontario for at least part of their life cycle. Some species are heavily dependent on wetlands and their adjacent habitats (Johnson, 1989). Riparian zones in other parts of North America have been identified as being important for amphibians and reptiles (Bury, 1988; Bury and Corn, 1988; Dickson, 1988). In pine forests in the southern U.S., Dickson (1988) found that amphibians and reptiles were abundant in streamside zones wider than 30 m. Apparently, streamside zones wide enough for a closed tree canopy, shaded understory, and a leaf litter ground cover provide adequate habitat for southern reptiles and amphibians.

Fish

Marshes and swamps provide spawning grounds, nursery habitat, and living areas for many important freshwater species. Vegetated buffer zones adjacent to wetlands will help protect these species. Buffer strips have been credited with being effective at protecting stream biota and habitat by maintaining shade and reducing sedimentation (Barton *et al.*, 1985; Brazier and Brown, 1973; Hartman *et al.*, 1987; Murphy *et al.*, 1986; Newbold *et al.*, 1980). The Ontario Ministry of Natural Resources has updated recommendations on the use of vegetative buffer strips to protect fish habitat (Anon., 1994). Vegetated buffer strips adjacent to wetlands should also help to protect fisheries resources present.

DISCUSSION

The only effective way of ensuring adequate protection of wetlands is to evaluate the site conditions and the likely effects of a proposed development on the wetland. From this base of information, a buffer width could be determined that would protect most, if not all, of the values. Where required, additional protective measures could be developed. Factors to be considered in determining the width of a buffer strip include land use and management above the strip, slope, width of slope above the strip, soil characteristics, type of vegetation, and type of wetland (marsh, swamp, bog, or fen).

Factors acting as constraints to implementation of effective buffer strips include cost of land and increased market value of land to the developer, increased tax value of developed property to the municipality, and inadequate time by agency staff to review individual proposals on a site-by-site basis to identify all relevant factors. In addition, pressure by developers, landowners, and others to minimize the width of the buffer strip, resistance by sectors of public to seeing land apparently "unused", lack of adequate knowledge to protect wetlands, and a tendency to seek the minimum width that could do the job are further constraints.

Options for identifying the appropriate width for the buffer strip on each site include an exhaustive review of all site factors, applying a formula to each site, establishing one value for each type of wetland, or establishing a minimum value for each type of wetland that is negotiable depending on site conditions and other considerations (e.g., deer wintering area, habitat of endangered or threatened species, waterfowl nesting habitat, slope, soil, nature of development, etc.).

As discussed, many factors should be considered in choosing a buffer width for each site. However, many factors operate to constrain this process. It is impractical for agency staff to thoroughly review all relevant factors on each site and apply a "tailor-made" buffer width to each situation. It is therefore necessary to establish a baseline buffer width for the protection of wetlands and then to make adjustments to this buffer width as appropriate. A brief review of the buffer widths used for protecting different natural resources is presented before providing recommended buffer widths.

To protect receiving streams from eroded sediment resulting from forestry practices, recommended buffer widths varied from 6 to 123 m, with the most common width being 30 m. In short-term agricultural experiments, grass filter strips from 4.1 to 90 m removed high percentages of sediment and sediment-bound nutrients. Up to 123 m of grassed filter strip was required for satisfactory removal of clay. Short-term experiments with buffer strips may underestimate widths needed for long-term effectiveness. In urban areas, for adequate removal of smaller-size sediment particles, buffer widths of 30.5 to 91.5 m are recommended.

Runoff in forested, agricultural, and urban areas tends to concentrate in rills and gullies which quickly moves water across slopes. The processes of settling, infiltration, and uptake by plants in the buffer strip are significantly reduced and their effectiveness is significantly reduced. Shallow uniform flow is required for effective removal of contaminants by buffer strips. Filter strip performance is best on slopes with a grade of 5% or less. A wider strip may guard against some of this deficiency. Filter strips will not function as intended for contaminant removal on slopes greater than 15%. With slopes greater than 15%, off-site runoff should be diverted around the filter strip rather than through it.

Quantitative data for widths of buffer strips or riparian zones required by different species and groups of wildlife are limited. Waterfowl require large blocks of nesting cover adjacent to wetlands in southern Ontario for reasonable nesting success. Female mallards may nest up to 300 m from the wetland edge.

Several species of songbirds in the midwestern states have been shown to nest successfully in streamside zones larger than 40 m. Bird species richness increases with the increasing width of wooded riparian habitats. Guidelines for forest management in Ontario recommend leaving 50 m uncut on either side of a stream or lake to provide for accipiters, buteos, eagles, cavity-nesting birds, and warblers. Riparian forests wider than 100 m are recommended for area-sensitive neotropical migrants.

Riparian areas are important to mammals as travel lanes, shelter areas, and feeding sites. In the southern U.S., streamside zones wider than 50 m in pine plantations maintained squirrel populations, but narrower streamside zones did not. In the same area, streamside zones wider than 30 m maintained populations of amphibians and reptiles.

Considering the information presented in this review, a baseline buffer width of 50 m is a reasonable distance to choose. With this width, many of the wetland functions on-site will receive some protection. This distance should be increased when on-site factors are such that wetland functions do not receive adequate protection.

From what has been presented, it is recommended that the required baseline buffer width be established as 50 m. The basic width of 50 m is to be used with increases based on limiting site conditions or special habitat requirements, e.g., steep slope, erodible soils, deer wintering area, red-shouldered hawk nesting area, or other relevant factor(s).

Marshes suitable for the production of waterfowl need considerably larger buffer widths. One of three options are recommended for marshes suitable for waterfowl production:

1. A buffer width of 300 m around the wetland, or
2. A buffer width of 50 m around the wetland plus an adjacent block of nesting cover greater than 12 ha within 500 m of the wetland (i.e., within reasonable walking distance of the wetland by newly hatched ducklings), or
3. At least 3 ha of buffer to 1 ha of wetland.

Bogs and fens are rare in southern Ontario and are sensitive to nutrient inputs (Riley, 1988). Buffer widths for these sites are to be established on a case-by-case basis. The basic minimum width is not likely to be adequate for protection and therefore an additional width of buffer should be determined for each site. Isolated bogs and fens in southern Ontario are small, usually with small catchment basins (Hooper, personal communication, 1994). The only way to ensure protection is to maintain the entire catchment basin in a naturally vegetated state. Those bogs and fens that occur within swamps and marshes in southern Ontario will be buffered to some degree by these surrounding wetlands. However, runoff from adjacent developments should be diverted away from that portion of the wetland containing the bog or fen.

Effectiveness of buffer strips must be monitored in a variety of situations to determine if they are protecting the values for which they are intended. Management of buffer strips must be revised based on the results of monitoring and accumulated knowledge, for which long-term studies are needed. Critical review of buffer strips should be updated and revised every 5 years based on new information from the scientific literature, results of monitoring, and management practices.

ACKNOWLEDGMENTS

The author expresses his appreciation to Marilyn Beatty for typing the paper.

REFERENCES

Allan, D. H. and Bohart, C. V., 1979. Soil conservation service programmes in non-game bird habitat management, in De Graaf, R. M. and Evans, K. E., eds., *Management of North Central and Northeastern Forests for Non-game Birds*, USDA For. Serv. Gen. Tech. Rept., NC-51, U.S. Department of Argriculture, Washington, D.C.

Anderson, W., 1957. A waterfowl nesting study in the Sacramento valley, California, 1955, *California Fish and Game* 43:71.

Anon., 1992a. *Wetlands Policy Statement*, A Statement of Ontario Government Policy issued under the authority of Section 3 of the Planning Act 1983, Ontario Ministry of Municipal Affairs and Ontario Ministry of Natural Resources, Queen's Park, Toronto.

Anon., 1992b. *Manual of Implementation Guidelines for the Wetlands Policy Statement*, Ontario Ministry of Municipal Affairs and Ontario Ministry of Natural Resources, Queen's Park, Toronto.

Anon., 1994. *Fish Habitat Protection Guidelines for Developing Areas*, Ontario Ministry of Natural Resources, Aquatic Ecosystems Branch, Willowdale.

Barker, S. E., 1983. The Development, Current Use and Effectiveness of Streamside Buffer Zones in Precluding Sediment Delivery to Forest Streams, M. Sc. Thesis., Department of Forestry, North Carolina State University, Raleigh.

Barton, D., Taylor, W., and Biette, R. M., 1985. Dimensions of riparian buffer strips required to maintain trout habitat in southern Ontario streams, *North American Journal of Fisheries Management* 5:364.

Bingham, S. C., Westerman, P. W., and Overcash, M. R., 1980. Effect of grass buffer zone length in reducing the pollution from land application areas, *Transactions of the ASAE* 23:330.

Blohm, R. J., Reynolds, A. E., Bladen, J. P., Nichols, J. D., Hines, J. E., Pollock, K. H., and Eberhardt, R. T., 1987. *Trans. N. Amer. Wildl and Nat. Res. Conf.* 52:246.

Brazier, J. R. and Brown, G. W., 1973. Buffer strips for stream temperature control, Paper 865, School of Forestry, Oregon State University, Corvallis, 97331.

Bury, R. B., 1988. Habitat relationships and ecological importance of amphibians and reptiles, in Raedeke, K. J., ed., *Streamside Management: Riparian Wildlife and Forestry Interactions*, University of Washington, Inst. of For. Res., Contr. No. 59: 61, Seattle.

Bury, R. B. and Corn, P.S., 1988. Responses of aquatic and streamside amphibians to timber harvest, in Raedeke, K. J., ed., *Streamside Management: Riparian Wildlife and Forestry Interactions*, University of Washington, Inst. of For. Res., Contr. No. 59:165 Seattle.

Clinnick, P. F., 1985. Buffer strip management in forest operations: a review, *Australian Forestry* 48(1):34.

Compton, B. B., Mackie, R. J., and Dusek, G. L., 1988. Factors influencing distribution of white-tailed deer in riparian habitats, *Journal of Wildlife Management* 52:544.

Corbett, E. S., Lynch, J. A., and Sopper, W. E., 1978. Timber harvesting practices and water quality in the eastern United States, *Journal of Forestry* 76(10):484.

Cowardin, L. M., Sargeant, A. B., and Duebbert, H. F., 1982. Problems and potentials for prairie ducks, *Naturalist* 34:4.

Cowardin, L. M., Gilmer, D. S., and Shaiffer, C. W., 1985. Mallard recruitment in the agricultural environment of North Dakota, *Wildlife Monographs* 92:37.

Croft, A. R. and Hoover, M. D., 1951. The relation of forests to our water supply, *Journal of Forestry* 49(4):245.

Dickey, E. C. and Vanderholm, D. H., 1981. Vegetative filter treatment of livestock feedlot runoff, *Journal of Wildlife Quality* 10(3):279.

Dickson, J. G., 1988. Streamside zones and wildlife in southern U.S. forests in *Practical Approaches to Riparian Resource Management: An Educational Workshop*, American Fisheries Society, Bethesda, MD.

Dillaha, T. A., Sherrard, J. H., Lee, D., Mostaghimi, S., and Shanholtz, V. O., 1985. Sediment and phosphorus transport in vegetative filter strips: phase I, field studies, *ASAE* Paper No. 85-2043.

Dillaha, T. A., Sherrard, J. H., Lee, D., Shanholtz, V. O., Mostaghimi, S., and Magette, W. L., 1986a. Use of vegetative filter strips to minimize sediment and phosphorous losses from feedlots: phase 1, experimental plot studies, VA *Water Res. Res. Ctr.*, Bull. 151, Backsburg, VA, 48.

Dillaha, T. A., Sherrard, J. H., and Lee, D., 1986b. Long-term effectiveness and maintenance of vegetative filter strips, VA *Water Res. Res. Ctr.*, Bull. 153, Backsburg, VA, 39.

Dillaha, T. A., 1989. Water quality impacts of vegetative filter strips, *ASAE* Paper No. 89-2043.

Dissmeyer, G. E. and Singer, J. R., 1977. Role of foresters in the area wide waste treatment management planning process, *So. J. Applied For.* 1(1):27.

Duebbert, H. F., 1969. High nest density and hatching success of ducks on South Dakota cap land, *Trans. North Am. Wildl. Nat. Resour. Conf.* 35:218.

Duebbert, H. F. and Lokemoen, J. T., 1976. Duck nesting in fields of undisturbed grass-legume cover, *J. Wildl. Manage.* 40(1):39.

Duebbert, H. F. and Kantrud, H. F., 1974. Upland duck nesting related to land use and predator reduction, *J. Wildl. Manage.* 38(2):257.

Evans, K. E. and Connor, R. N., 1979. Snag management, in DeGraaf, R. M. and Evans, K. E., eds., *Management of North Central and Northeastern Forests for Non-game Birds*, USDA, For. Ser., General Technical Report NC-51, U.S. Department of Agriculture, Washington, D.C.

Glooschenko, V. and Grondin, P., 1988. Wetlands of eastern temperate Canada in Wetlands of Canada, *Ecol. Land Class.* Ser. #24., Sustainable Development Branch, Environment Canada, Ottawa, Ont., chap. 6.

Greenwood, R. J., Sargeant, A. B., Johnson, D. H., Cowardin, L. M., and Terry Shaffer, T.L., 1987. Mallard nest success and recruitment in prairie Canada, *Trans. North Am. Wildl. and Nat. Res. Conf.* 52:298.

Hartman, G. J., Scrivener, C., Holtby, L. B., and Powell, L., 1987. Some effects of different streamside treatments on physical conditions and fish population processes in Carnation Creek, a coastal rain forest stream in British Columbia, in Salo, E. O. and Cundy, T. W., eds., *Streamside Management: Forestry and Fishery Interactions*, Contr. No. 57, Institute For. Resour., University of Washington, Seattle.

Haupt, H. F. and Kidd, J. W., Jr., 1965. Good logging practices reduce sedimentation in central Idaho, *J. For.* 63(9):664.

Higgins, K. F., 1977. Duck nesting in intensively farmed areas of north Dakota. *J. Wildl. Manage.* 41:232.

Hochbaum, G. S., Caswell, F. D., Turner, B. C., and Nesman, D. J., 1987. Relationships among social components of duck breeding populations, production and habitat conditions in prairie Canada, *Trans. N. Amer. Wildl. and Nat. Res. Conf.* 52:310.

Hooper, G., personal communication, 1994.

Hornbeck, J. W. and Reionhart, K. G., Water quality and soil erosion as affected by logging in steep terrain, *J. of Soil and Water Cons.*, 19:23.

James, R. D., 1984a. Habitat management guidelines for Ontario's forest nesting accipiters, buteos and eagles, Wildlife Branch, Ontario Ministry of Natural Resources, Toronto.

James, R. D., 1984b. Habitat management guidelines for cavity nesting birds in Ontario, Wildlife Branch, Ontario Ministry of Natural Resources, Toronto.

James, R. D., 1984c. Habitat management guidelines for warblers of Ontario's northern coniferous forests, mixed forests or southern hardwood forests. Wildlife Branch, Ontario Ministry of Natural Resources, Toronto.

Jarvis, R. L. and Harris, S. W., 1971. Land-use patterns and duck production at Malheir National Wildlife Refuge., *J. Wildl. Manage.* 35:767.

Johnson, B., 1989. Familiar Amphibians and Reptiles of Ontario, Natural Heritage/Natural History, Inc., P.O. Box 69, Postal Station H., Toronto, 168.

Johnson, D. H. and Sargeant, A. B., 1977. Impact of red fox predation on the sex ratio of prairie mallards, *U.S. Fish Wildl. Serv.*, Wildl. Res. Rep. 6, 56, Washington, D.C.

Kalmbach, E. R., 1939. Nesting success: its significance in waterfowl production. *Trans. North Am. Wildl. Conf.* 4:591.

Kauffman, J. B., 1988. The status of riparian habitats in Pacific northwest forests, in Streamside Management: Riparian Wildlife and Forestry Interactions, Raedeke, K. J., ed., University of Washington, Inst. of For. Res., Contribution No. 59, Seattle.

Keith, L. B., 1961. A study of waterfowl ecology on small impoundments in southeastern Alberta, *Wildlife Monographs* No. 6.

Kellar, C. M. E., Robbins, C. S., and Hatfield, J. S., 1993. Avian communities in riparian forests of different widths in Maryland and Delaware, *Wetlands* 13(2):137.

Labisky, R. F., 1957. Relation of hay harvesting to duck nesting under a refuge-permittee system, *J. Wildl. Manage.* 21:194.

Lammers-Helps, H. and Robinson, D. M., 1991. Literature review and summary pertaining to buffer strips, Soil and Water Cons. Info. Bur., University of Guelph, Guelph, Ontario, 46.

Livezey, B. C., 1981. Locations and success of duck nests evaluated through discriminant analysis, *Wildfowl* 32:23.

Magette, W. L., Brinsfield, R. B., Palmer, R. E., and Wood, J. D., 1986. Vegetated filter strips for non-point source pollution control: nutrient considerations, *ASAE* Paper No. 86-2024.

MacArthur, R. H., MacArthur, J. W., and Preer, J., 1962. On bird species diversity. II. Prediction of bird census from habitat measurements, *Am. Natur.* 96:167.

Murphy, M. L., Heifetz, J., Johnson, S. W., Koski, K.V., and Thedings, J. K., 1986. Effects of clear-cut logging with and without buffer strips on juvenile salmonids in Alaskan streams, *Can. J. Fish. Aquat. Sci.* 43:1521.

Nelson, H. K. and Duebbert, H. F., 1973. New concepts regarding the production of waterfowl and other game birds in areas of diversified agriculture, *Int. Congr. Game Biol.*, Stockholm, Sweden, XI. 18.

Newbold, J. D., Erman, D. G., and Roby, I. K. B., 1980. Effects of logging on macroinvertebrates in streams with and without buffer strips, *Can. J. Fish. Aquat. Sci.* 37:1076.

Oakley, A. L., Collins, J. A., Everson, L. B., Heller, D. A., Howerton, J. A., and Vincent, R. E., 1985. Riparian zones and freshwater wetlands, in Management of Wildlife and Fish Habitats of Western Oregon and Washington, E. R. Brown, ed., USDA For. Serv., Rep. R6-F&WL-192-1985, U.S. Department of Agriculture, Washington, D.C.

Occoquan Watershed Monitoring Laboratory (OWML), 1987. Final contract report: Washington area NURP project prepared for metropolitan Washington council of governments, Manassas, Virginia, 1983, 450.

Packer, P. E., 1967. Criteria for designing and locating logging roads to control sediment, *Forest Science* 13(1):2.

Paterson, J. J., Jones, J. H., Olsen, F. J., and McCoy, G. C., 1980. Dairy Liquid waste distribution in an overland flow vegetative-soil filter system, *Transactions of the ASAE* 23:973.

Raedeke, K. J., Taber, R. D., and Parge, D. K., 1988. Ecology of large mammals in riparian systems of Pacific northwest forests, in Raedeke, K. J., ed., *Streamside Management: Riparian Wildlife and Forestry Interactions*, University of Washington, Inst. of For. Res., Contribution No. 59:275, Seattle.

Riley, J. L., 1988. Southern Ontario bogs and fens of the Canadian shield, in *Wetlands: Inertia or Momentum, Proceedings of a Conference*, Toronto, Ontario, Oct. 21–22, 1988, Fed. of Ont. Nat., Don Mills, Ontario.

Schueler, T. R., 1987. Controlling Urban Runoff, *A Practical Manual for Planning and Designing Urban BMP's*, Wash. Metr. Water Res. Plan. Bd., Washington, D.C.

Shisler, J. K., Waidelich, P. E., and Russell, H. G., 1987. Buffer zones in wetland management practice in *Estuarine and Coastal Management — Tools of the Trade, Proceedings of the Tenth National Conf. of the Coastal Society*, New Orleans, LA.

Smith, R. H., 1953. A study of waterfowl production on artificial reservoirs in eastern Montana, *J. Wildl. Manage.* 17:276.

Snell, E. A., 1987. Wetland distribution and conversion in southern Ontario, Environment Canada, Working Paper No. 48, 53, Toronto.

Sparrow, R. D. and Patterson, J. H., 1987. Conclusions and recommendations from studies under stabilized duck hunting regulations, management implications and future directions, *Trans. N. Amer. Wildl. and Nat. Res. Conf.* 52:320.

Stauffer, D. F. and Best, L. B., 1980. Habitat selection by birds of riparian communities: evaluating effects of habitat alterations, *J. Wildl. Manage.* 44:1.

Thomas, J. W., Maser, C., and Rodiek, J. E., 1979. Riparian Zones, in Thomas, J. W., ed., *Wildlife Habitats in Managed Forests, The Blue Mountains of Oregon and Washington*, USDA For. Serv. Agr., Handbook No. 553, U.S. Department of Agriculture, Washington D.C., chap. 3.

Trimble, G. R., Jr. and Sartz, R. S., 1957. How far from a stream should a logging road be located? *J. For.* 55(5):339.

Turner, B. C., Hochbaum, G. S., Caswell F. D., and Nieman, D. J., 1987. Agricultural impacts on wetland habitats on the Canadian prairies, 1981–85. *Trans. N. Amer. Wildl. and Nat. Res. Conf.* 52:206.

van Groenewoud, H., 1977. Interim recommendation for the use of buffer strips for the protection of small streams in the Maritimes, Info. Rept. M-X-74, Can. For. Serv., Maritimes For. Res. Ctr., Fredericton, N.B.

Voigt, D. R., 1990. Timber Management Guidelines for the Provision of White-tailed Deer Habitat, Ontario Ministry of Natural Resources, Toronto.

Wilson, L. G., 1967. Sediment removal from flood water by grass filtration, *Transactions of the ASAE*, 10(1).

Young, R. A., Huntrods, T., and Anderson, W., 1980. Effectiveness of vegetated buffer strips in controlling pollution from feedlot runoff, *J. Environ. Quality* (9)3:483.

19   SUMMARY OF FINAL SESSION

The two-day symposium concluded with a panel discussion. The panel discussion concentrated on the different potential areas of wetland priority research. Summarized below are the important recommendations related to the symposium theme and some research needs, especially in the area of general wetland management.

1. There is a major concern regarding the lack of support for scientists interested in taxonomy, systematics, and natural history in both the U.S. and Canada. These branches of biology are basic to any ecological study on wetlands, with plant taxonomy being fundamental to boundary delineation. The assessment of species diversity in wetlands is important in wetland evaluation; and the field of taxonomy must receive special status in wetland research. Further studies on the common and dominant wetland ecotypes are recommended.

2. There is a lack of clear understanding of the linkages among the plant-soil-hydrology triune. Correlation studies of these three components, especially in a wetland-upland gradient, are needed.

3. There are major limitations in existing methods used to delineate boundaries that have evolved during the past 30 years. The Primary Indicators Method, as discussed in Chapter 8, represents an important step in the right direction for developing an appropriate boundary delineation method. Adjustments to the method, however, may have to be developed in response to whether one is discussing mapping wetlands in the field as opposed to mapping a wetland at 1:24,000 scale. As well, the need exists to have the methods adjustable if the delineation of a wetland boundary is a "tax mapping", where one tenth of a hectare can make a significant difference in somebody's tax bill vs. mapping wetlands at the regional scale or at the watershed or county scale.

4. For establishing necessary buffer widths, selecting an appropriate method depends on the scale. For example, methods for developing buffer guidelines differ for wetlands at the watershed or site-specific scales. The CREAMS Model, for establishing buffers (Chapter 17), is a suitable model at a fairly local scale, but at the watershed scale an appropriate model is not so well developed. An important point of discussion is whether there is a need to set minimum and maximum buffer widths. This leads to the question of considering wetland functional boundaries and the necessity to examine such boundaries for establishing buffers.

1-56670-147-3/96/$0.00+$.50
© 1996 by CRC Press, Inc.

5. Regarding ecotones and their role in wetland ecology, Dr. Holland's specific recommendations on research needs are as follows:

Study the importance of upland/wetland ecotones in a variety of landscapes simultaneously.

Characterize the relationship between wetland size, hydrologic characteristics, and dimensions of the ecotone on the assimilative capacity of the wetland patches and ecotones.

Use existing management questions to develop a series of experiments that will test our ability to successfully maintain or enhance the functions of wetland ecotones.

Identify traditional, low-intensity management techniques that have successfully maintained or enhanced the functions of wetland ecotones in the past.

Utilize existing descriptive and predictive models to identify parameters of wetland patches and ecotones that need to be better understood.

Establish lateral transects crossing from uplands to wetlands to open water ecosystems and assess biological diversity in wetland/upland and wetland/open water ecotones.

The following recommendations for research relate to the management of wetland patches and ecotones. The wise use of wetland patches and ecotones requires action on a large scale, where consideration is given to all factors affecting wetlands and the drainage basins of which they are a part. A good fundamental understanding of ecosystem and hydrological processes is necessary for good management. Careful emphasis and integration might ultimately result in national wetland policies which include consideration of upland/wetland and wetland/open water ecotones. The major items for developing such policies may include:

- Need for a national inventory of wetland patches and ecotones
- Need to identify the benefit and values of these wetland patches and ecotones
- Need to define the priorities for each site in accordance with the needs of socioeconomic conditions
- Require a proper assessment of environmental impact before development projects are approved, a continuing evaluation during the execution of projects, and the full implementation of environmental conservation measures which take full account of the recommendations of this process of environmental assessment and evaluation
- Use of development funds for projects which permit conservation and sustainable utilization of wetland resources

6. A strong concensus was apparent on the necessity for long-term studies on wetlands. The discontinuation of some of the long-term studies was deprecated by all members attending the symposium. One classic example, quoted from Ontario experience, was the case of the Wally Creek Drainage Project near Cochrane, a multimillion dollar project which was discontinued. Similarly, the Irish Creek Wetland study in southern Ontario is having difficulty to sustain the

research activity. There was a plea to combine long-term wetland studies with the establishment of reference wetlands in the U.S. and Canada. Such reference wetlands could be used for accumulation of detailed ecological data. It was also suggested that the reference wetlands should be fairly representative and not necessarily "unique areas" which may not have transfer value. An example of a reference wetland was a USGS site in Minnesota under a program called Interdisciplinary Research Initiatives.

7. A popular recommendation was the need to encourage and establish interdisciplinary studies. There were serious reservations on the ongoing piecemeal research where separate wetlands are examined for hydrology, soils, fauna, and flora. To undertake interdisciplinary studies, it is imperative that scientists willing to share and integrate results are rewarded in some way. The focus should be in developing more meaningful and useful products from an environmental management perspective and not just in discussing a single feature such as "birds" in isolation. An example from The Netherlands that was quoted was the *Plantago* multidisciplinary study financially supported by the Government of the Netherlands and the European Community. The project was started five years ago with the cooperation of six universities and included genetic, physiological, ecological, landscape, and nature management approaches.

8. Several participants questioned the relevancy of wetland research and its usefulness to others. An interesting example from the U.S. illustrated this point. Regarding the National Acid Precipitation Assessment Program, the U.S. Congress in fact at one time asked for "assessment" and not for the production of tons of reports. What was questioned were the costs and benefits of such programs. It is also important that the results of scientific investigations on wetlands be presented in such a way that they are understood by a wider audience.

9. Another aspect that needs immediate attention is to combine wetland research efforts with wetland planning and management. It is important to understand how land uses modify and alter wetland functions and values. Further, development of mitigative measures of land use impacts should receive immediate attention. At the present time there are frequent conflicts and controversies on urban development in and around significant wetlands. This involves regulatory agencies and public institutions like the university on one side and the development industry on the other. An important constituency that is generally excluded is the public/citizens who play a major role in educating society. A positive change in societal attitudes towards the natural environment in general, and a concern for wetlands in particular, is expected to moderate the ongoing strife with the development industry and those agencies with the mandate of protection and conservation.

10. The concluding recommendation is for wetland groups in different regions of the world to improve networking for exchange of information and to achieve international cooperation and collaboration.

INDEX

A

Accipiter cooperii, 168
Accipiters, 168, 271, 273
Acer, 196, 225
  A. *campestre,* 95
  A. *pseudoplat.,* 95
  A. *rubrum,* 12, 118, 120
Achillea
  A. *millifolium,* 225
  A. *ptarmica,* 97
Acrotelm, 47–48
Adaptive mechanisms, in river floodplains,
    91–107, 118
Administrative boundaries, 139
Adsorption of pollutants, 70
Aerenchyma, 91, 100, 102–103, 106
  schizogenous, 100
  shoot elongation and, 103–104
Aerial photography, 127, 177–178
  data analysis, 180–182
  Presque Isle, Pennsylvania, 179–180
  variability of measurements, 182–183
Aerobic soil, 27, 31
Agelaius phoeniceus, 169
Aggregate extraction
  activities during, 241–242
  mitigation and separation strategies,
    246–248
  scenarios, 245–246
  types of, 245
Aggregate extraction sites, 237–248
  Ontario, percent in sensitive areas, 248
Aggregate Resources Act, 237–239,
    246
Agricultural Research Centre of Finland,
    223
Agricultural Research Service, 253
Agricultural runoff, 218, 252
  buffer strips used for, 266–268
  nitrogen, 10–11, 221–222
  pesticides, 210
  phosphorous, 33, 210, 221–222,
    267– 268
  sediment load of, 266–268

Agriculture, 1, 61, 165
  Finland, 221–222
  grazing, 91–92, 96–97, 106, 218
  haymaking, 96–97
  mowing, 97, 106, 233
  New York, central, 196
Agrostis
  A. *stolonifera,* 97
  A. *tenuis,* 225
Aix sponsa, 169
Alaska floodplains, 105
Alberta, land use planning of Edmonton,
    139–140, 143–149
Alder, 96, 225
Alisma, 120
Alligatorweed, 120
Alnus, 225
  A. *incana,* 96
Alopecurus pratensis, 97
Alternantera philoxeroides, 120
Amblystegiaceae, 45
Ambrosia, 49. See also Pollen
  degraded bogs, 60
  floating mats, 54
American Frog-bit, 120
Ammonium, 226, 232
Amphibians, 142, 271–272. See also Wildlife
Anaerobic soil, 27, 29, 80, 102–103, 117
Andromeda glaucophylla, 120
Annual life history, 98
Anoxic layer of bogs (catotelm), 47–48
Anthropogenic factors, 20–22, 132, 218
  artificial wetlands, 2, 62
  classification of impact on peatlands, 61–63
  European settlement in peatlands, 47,
    52–61
  floodplain grasslands, 96–98
  river floodplains, 93
  succession and, 156–157
  urban intensification and environmental
    stability, 67–76
Apalachicola floodplain wetlands, 14–15
*Application of Satellite Data for Mapping and
    Monitoring Wetlands - Fact Finding
    Report,* 127